Biological and Medical Physics, Biomedical Engineering

For other titles published in this series, go to
www.springer.com/series/3740

Roger Narayan · Thomas Boland
Yuan-Shin Lee
Editors

Printed Biomaterials

Novel Processing and Modeling Techniques for Medicine and Surgery

Editors
Roger Narayan
Department of Biomedical Engineering
University of North Carolina
Chapel Hill, NC
USA
roger_narayan@unc.edu

Yuan-Shin Lee
Edward P. Fitts Department of Industrial and Systems Engineering
North Carolina State University
Raleigh, NC
USA
yslee@ncsu.edu

Thomas Boland, PhD
Professor
Department of Metallurgy and Materials Engineering
The University of Texas at El Paso
El Paso, TX 79968-0520
tboland@utep.edu

ISBN 978-1-4419-1394-4 e-ISBN 978-1-4419-1395-1
DOI 10.1007/978-1-4419-1395-1
Springer New York Dordrecht Heidelberg London

Library of Congress Control Number: 2009940802

Printed on acid-free paper

Springer is part of Springer Science+Business Media (www.springer.com)

Preface

The creation of substitutes to repair damaged tissues and organs dates back to the beginning of recorded history [1]. Several ancient civilizations dabbled in tissue repair; for example, Indian physicians created primitive skin grafts as long ago as 800 BC. It has been only within the past century that surgical understanding of vessel anastamosis and aseptic surgical technique have enabled transplantation and replacement of tissues [2].

There are many techniques for harvesting natural tissue for transplant use. The "gold standard" for natural transplantable tissue is called autograft tissue. This type of tissue is transferred from one site to another in the same individual. If one is lucky enough to have a genetically identical twin, also known as a monozygotic twin, this individual can serve as a source of isograft tissue. Autograft/isograft tissue use is associated with many problems. For example, additional surgery at the "donor" site can result in complications, including infection, inflammation, and chronic pain. In addition, the quantity of material that can be harvested from the donor site is limited.

Another source of transplantable tissue is known as allograft tissue. In this case, tissue is transferred from one person to another. Over 20 different types of tissue, including cartilage, cornea, hearts, kidney, liver, lung, and pancreas, have been successfully transplanted between different individuals. Unfortunately, this type of tissue use is also associated with many difficulties. The most significiant problem with allograft transplantation is providing an adequate amount of organs for all of the patients who need them. There are currently over 80,000 people on waiting lists for allograft transplantation in the United States [3]. Because of this supply limitation, more than 10,000 people have died in recent years on waiting lists for allograft organs and tissues. In addition, the body's immune system generates acute vascular rejection and chronic rejection processes that degrade transplanted material in days, weeks, months, and years after implantation [4]. The long-term immunosuppressive therapy typically used to counter the rejection process may itself lead to tumor formation.

There is also a risk of infectious disease transmission from the allograft donor to the allograft recipient [5]. Although allograft tissue may be treated using gamma irradiation, electron beam radiation, freeze-drying, ethylene oxide, or tissue freezing methods, the risk of disease transmission persists [6]. In addition, many methods that are used to reduce disease transmission also decrease viability of the tissue.

The risks of infection after transplantation of allograft tissue are not theoretical. For example, in November 2001, a 23-year-old otherwise healthy Minnesota man died from an Clostridium sordellii infection after undergoing transplantation of allograft femoral condyle tissue. That Food and Drug Administration (FDA) and the Centers for Disease Control and Prevention (CDC) traced the allograft tissue to a commercial tissue bank, CryoLife, Inc., in Kennesaw, Georgia. CryoLife was ordered to recall its allograft tissue and was temporarily shut down by the FDA. The CDC then asked orthopedic surgeons to report infections associated with allograft transplantations. The CDC identified 26 allograft-related infections, 11 of which were Clostridium septicum or Clostridium sordelli infections that involved allograft tissue processed by CryoLife. It has been estimated that the risk of HIV transmission with allograft bone is one case in 1.6 million. Similarly, one case of hepatitis B transmission and three cases of hepatitis C transmission have been clinically correlated with allograft tissue transplantation. Xenografts, or grafts from animals are rare, as these grafts allow transfer of animal pathogens (bacteria, viruses, fungi, and prions) to humans.

The growing demand for tissue substitutes and the continuing limitations of natural tissue substitutes have led to the development of a field known as tissue engineering. This field was pioneered by Robert S. Langer, Joseph P. Vacanti, and Anthony Atala at the Massachusetts Institute of Technology and Harvard University. The materials used in tissue engineering include living cells, natural materials, and synthetic materials. Tissue engineered materials are created by placing living cells within scaffolding that is meant to guide cell growth, differentiation, and development. The cell-seeded structure is then placed in a bioreactor that provides oxygen and nutrients, which enables cells to multiply within the scaffold. The tissue substitute is then implanted in an environment that will permit the tissue to possess normal structure and/or exhibit normal function.

Current tissue engineering processing techniques have yet to overcome several limitations. First, cell division is not rapid and the scaffold seeding process is difficult. In addition, it is very difficult to create tissue substitutes that contain more than several cell layers because bioreactors cannot provide sufficient nutrients to thicker structures. Growth in a bioreactor usually ceases after the tissue is 100 μm thick. These problems have severely limited the clinical use of tissue substitutes fabricated using conventional methods. As a result, only been a handful of tissue substitutes created using conventional tissue engineering methods have been approved by the FDA for use in the United States.

Several investigators have recently examined the use of rapid prototyping technologies to overcome the limitations associated with current tissue engineering processing methods. This technology was originally developed over a quarter century ago for the fabrication of prototypes of machine tools, automotive parts, and military devices. The term "rapid prototyping" is used to describe the fabrication of three-dimensional structures through additive joining of materials in a layer-by-layer manner as opposed to conventional subtractive processes. Recent studies have shown that printing techniques and other rapid prototyping methods may be used to process cells and scaffold materials in order to create patient-specific tissue substitutes. Data obtained from magnetic resonance imaging or computed tomography of a given patient may be employed in order to create models of the injured, damaged, or missing tissue. Customized implants created using printing techniques may possess suitable features geometry, weight, and biological properties for treatment of a certain patient. Surface features may be incorporated into prostheses in order to increase diffusion of nutrients to cells on the prosthesis surface and promote desirable tissue-implant interactions. In addition, many rapid prototyping technologies can be placed near clinical facilities; specialized or dedicated fabrication environments are typically not required. The typical feature size, advantages, and disadvantages for common rapid prototyping and printing techniques are provided in Table 1.

Patients and surgeons are demanding more individualized, "patient-specific" treatments for trauma, injury, aging, and disease processes. The printing technologies described in this volume offer tremendous potential for the fabrication of tissue substitutes with appropriate mechanical and biological properties for treatment of a given patient. We anticipate that the use of printed biomaterials in medicine, surgery, and dentistry will become more significant in the next several years.

Chapel Hill, NC — Roger J Narayan
El Paso, TX — Thomas Boland
Raleigh, NC — Yuan-Shin Lee

References

1. Converse JM, Casson PR (1968) The historical background of transplantation. In: Rapaport FT, and Dausset J (ed) Human Transplantation. Grune & Stratton, New York
2. Carrel A (1905) The transplantation of veins and organs. Am Med 10:1101–1102
3. 2000 Annual Report of the U.S. Scientific Registry of Transplant Recipients and the Organ Procurement and Transplantation Network: Transplant Data 1989–1998 (2001, February 16). HHS/HRSA/OSP/DOT and UNOS, Rockville, MD and Richmond, VA
4. Stock, UA, Vacanti JP (2001) Tissue engineering: current state and prospects. Annu Rev Med 52:443–451
5. Charlton B, Auchincloss H, Fathman CG (1994) Mechanisms of transplantation tolerance. Annu Rev Immunol 12:707
6. Boyce T, Edwards J, Scarborough N (1999) Allograft bone. The influence of processing on safety and performance. Orthop Clin North Am 30(4):571–81

Table 1 Rapid prototyping and printing technologies used for additive processing of biomaterials

Technique	Feature size (μm)	Advantages	Disadvantages
Rapid prototyping robotic dispensing system (RPBOD)	400–1,000	Compatible with many materials; Biological molecules may be included	Precise control of precursor material properties essential; Freeze drying required
Robocasting	100–1,000	Compatible with many materials	Precise control of precursor material properties essential
Selective laser sintering (SLS)	500	Microporous structures may be produced; Compatible with several materials; Rapid processing rate	Precursor material must be in powder form; High processing temperatures involved; Powdery surface finish; Completed part may contain trapped powder
Precise extrusion manufacturing (PEM)	200–500	Precursor material must be in pellet form	High processing temperatures involved; Difficult to prepare structures with microscale porosity
Low-temperature deposition manufacturing (LDM)	400	Biological molecules may be included	Use of solvent required; Freeze drying required
Multi-nozzle deposition manufacturing (MDM)	400	Compatible with several materials; Biological molecules may be included; Low processing temperatures involved	Use of solvent required; Freeze drying required
TheriForm™	300	Microporous structures may be produced; Compatible with many materials; Rapid processing times	Precursor material must be in powder form; Powdery surface finish; Completed part may contain trapped powder
3D Bioplotter	250	Compatible with several biomaterials; Biological molecules may be included	Low mechanical strength; Low accuracy; Slow processing rate
3D Fiber-deposition technique	250	Precursor material must be in pellet form	High processing temperatures involved; Difficult to prepare structures with microscale porosity
Fused deposition modeling (FDM)	250	Good mechanical strength; Good control of internal microstructure; Good control of external microstructure	High processing temperatures involved; Filament precursor material; Difficult to prepare structures with microscale porosity
Stereolithography apparatus (SLA)	250	Compatible with many materials; Rapid processing rate	Material must be biocompatible and capable of photopolymerization; Requires use of ultraviolet light

3-dimensional printing™	200	Compatible with several materials; Microporous structures may be produced; Water may be used as a binder; Rapid processing rate	Precursor material must be in powder form; Completed part may contain trapped powder; Powdery surface finish; Post processing steps usually required; Low mechanical strength;
3-D inkjet printer	180	Compatible with several materials; Good control of internal microstructure; Good control of external microstructure	Multiple processing steps necessary
Two photon polymerization	<1	Control of external and internal morphology; Requires use of infrared light	Material must be biocompatible and capable of photopolymerization

Modified from "Rapid Prototyping of Artificial Tissues and Medical Devices," Boland, Thomas; Ovsianikov, Aleksandr; Chickov, Boris N.; Doraiswamy, Anand; Narayan, Roger J.; Wai Yee Yeong; Kah Fai Leong; Chee Kai Chua, Advanced Materials & Processes, 165(4):51–53, 2007.

Contents

Contributors

Betsy K. Davis, DMD, MS
Director, Maxillofacial Prosthodontic Clinic, Associate Professor, Department of Otolaryngology and Head & Neck Surgery and Oral and Maxillofacial Surgery, Medical University of South Carolina, 135 Rutledge Avenue, Charleston SC 29425, USA

Salil Desai
Department of Industrial and Systems Engineering, North Carolina A&T State University, Greensboro, NC 27411, USA

Randy Emert
Coordinator of the Engineering Graphics Program, Lecturer, Engineering Graphics, Clemson University, M-11 Holtzendorff Hall, Clemson University, Clemson, SC 29634, USA

S. Güçeri
Laboratory for Computer-Aided Tissue Engineering, Department of Mechanical Engineering and Mechanics, Drexel University, Philadelphia, PA 19104, USA

Benjamin Harrison
Wake Forest Institute for Regenerative Medicine, 391 Technology Way, Winston-Salem, NC, USA

Yuan-Shin Lee
Edward P. Fitts Department of Industrial and Systems Engineering, North Carolina State University, Raleigh, NC 27695, USA

Shiyong Lin
Edward P. Fitts Department of Industrial and Systems Engineering, North Carolina State University, Raleigh, NC 27695, USA

Roger J. Narayan
Department of Biomedical Engineering, University of North Carolina, Chapel Hill, NC 27599, USA

L. Shor
Laboratory for Computer-Aided Tissue Engineering, Department of Mechanical Engineering and Mechanics, Drexel University, Philadelphia, PA 19104, USA

Binil Starly
School of Industrial Engineering, University of Oklahoma, Norman, OK 73019, USA

W. Sun
Laboratory for Computer-Aided Tissue Engineering, Department of Mechanical Engineering and Mechanics, Drexel University, Philadelphia, PA 19104, USA

Tao Xu
Wake Forest Institute for Regenerative Medicine, 391 Technology Way, Winston-Salem, NC 27157, USA
College of Engineering, University of Texas at El Paso, El Paso, TX 79968, USA

E.D. Yildirim
Laboratory for Computer-Aided Tissue Engineering, Department of Mechanical Engineering and Mechanics, Drexel University, Philadelphia, PA 19104, USA

James J. Yoo
Wake Forest Institute for Regenerative Medicine, 391 Technology Way, Winston-Salem, NC 27157, USA

Yuyu Y. Yuan
Department of Bioengineering, Clemson University, Clemson, SC 29634, USA

Chapter 1
Surgical Cutting Simulation and Topology Refinement of Bio-Tissues and Bio-Object

Shiyong Lin, Yuan-Shin Lee, and Roger J. Narayan

Abstract In this paper, we propose methodology and algorithms to generate realistic cuts on heterogeneous deformable object models. A three-dimensional node snapping algorithm is presented to modify the topology of deformable models, without adding new elements. Smooth cut is generated by duplicating and displacing mass points that have been snapped along the cutting path. Several sets of triangles representing different soft tissues are generated along the new cut to present the internal structures and material properties of heterogeneous deformable objects. A haptic device is integrated into cutting simulation system as a cutting tool. The proposed cutting techniques can be used in surgical simulation or other virtual simulations involving topological modification of heterogeneous soft materials to enhance the fidelity and realism.

1.1 Introduction

Medical surgical simulation is a technology dedicated to medical training and surgery planning. Cutting simulation is the key component and also the most difficult one in surgical simulators. Cutting is a common operation in both conventional and minimal invasive surgeries. Most surgical tasks begin with an incision to expose the surgical region. However, it is a difficult task to simulate cutting operation since the topology of deformable models is changing in real-time and a large amount of computation is required. Cutting in real surgery is a careful step-by-step procedure to reach desirable cutting depth. The cutting tool is very often perpendicular to the target surface. Each movement of cutting tool is relatively simple, often short and straight [1]. Soft tissues

S. Lin and Y-S. Lee (✉)
Edward P. Fitts Department of Industrial and Systems Engineering,
North Carolina State University, Raleigh, NC, 27695, USA
e-mail: yslee@ncsu.edu

R.J. Narayan
Department of Biomedical Engineering, University of North Carolina, Chapel Hill, NC, 27599, USA

R. Narayan et al. (eds.), *Printed Biomaterials*, Biological and Medical Physics, Biomedical Engineering, DOI 10.1007/978-1-4419-1395-1_1,

are typically not cut too deep at once for the sake of avoiding unnecessary damage to internal healthy tissues. For example, in the middle line laparotomy, surgical incision routine begins with skin incision, followed by subcutaneous layer incision and peritoneum incision [2]. It is very important to show realistic internal structures at the cutting site as the visual clue, especially for surgical training and planning purposes.

In most of the current cutting simulations, cutting operation is considered only with surface models or homogeneous volumetric models [3–5]. In general, it is relatively easy to modify the topology of surface-based models since only surface meshes are involved in cutting. Groove or gutter is often created along the cutting path to give the realistic illusion of volumetric cut [6, 7]. Surface-based models are not suitable for most soft tissues like muscles, especially when internal structures or material properties of these models play a critical role in the cutting procedure.

On the other hand, cutting simulation on volumetric deformable models is normally implemented by three types of topological modification approaches, e.g., element removal, mesh subdivision, and mesh adaptation. Element removal is directly removing an element intersecting with the tool [8]. Despite of its simplicity and computational efficiency, this method cannot present smooth cut because of visual artifacts. Mesh subdivision can generate smooth cutting path by subdividing tetrahedrons [3] along the cutting planes. However, the method is computationally expensive due to many subdivision cases and increasing element number. Another drawback is that smaller or degenerated elements may be created near the cutting site, which can cause instability of the simulation [3]. Mesh adaptation (or node snapping) shown can generate good cutting path without creating new elements. Mass points near cutting path are snapped to the closest points on the cutting path such that some related edges (or springs) are aligned along the cutting path. Mesh adaptation was applied in tetrahedral mesh [9]. In our previous work, heterogeneous deformable object modeling technique has been proposed to model heterogeneous soft tissues [10, 11]. The quality of resulting incision is limited by initial mesh resolution. The smaller size of elements, the smoother cut, can be accomplished by node snapping. However, current cutting approaches only deal with homogeneous tetrahedral models. Soft tissue layers around the cutting site couldn't be presented as often observed in real surgery.

In this paper, a 3D cutting method is proposed to simulate the medical cutting operation on heterogeneous deformable models. Technique of constructing heterogeneous deformable models is presented for modeling heterogeneous biological soft tissues. To simulate surgical cutting, a 3D topological modification method is presented for cutting simulation on heterogeneous deformable models. The presented techniques can be used for medical surgical simulation and deformable objects modeling.

1.2 Heterogeneous Deformable Object Modeling

Biological soft tissues are viscoelastic, anisotropic, and heterogeneous deformable objects [12]. As shown in Fig. 1.1, a heterogeneous deformable object may include various soft and hard materials. For example, human legs consist of skin, fat tissue, various muscles, bones, and other soft tissues. It is of great importance to model biological soft tissues heterogeneously to simulate realistic surgical procedures [10]. We have proposed a tri-ray node snapping algorithm to generate heterogeneous deformable models from boundary or interface surfaces of biological soft tissues [13]. Figure 1.1 shows the boundary surfaces of skin, bones, and various muscles in a human leg and also the heterogeneous human leg model generated from these boundary surfaces.

Deformable object model generation is a process of discretizing volumetric objects into mass points and springs in Mass Spring Model or small elements in Finite Element Model [14, 15]. In our previous work in [13], heterogeneous deformable models of biological soft tissues have been generated based on volumetric Mass Spring Model. The heterogeneous deformable model consists of hexahedral mass spring elements, as shown in Fig. 1.2. Mass points on the same layer are connected

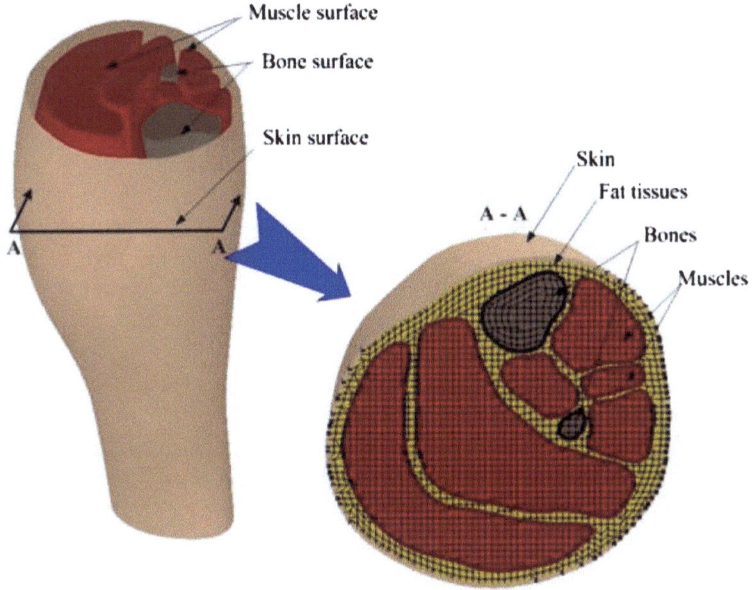

Fig. 1.1 Heterogeneous soft tissue model

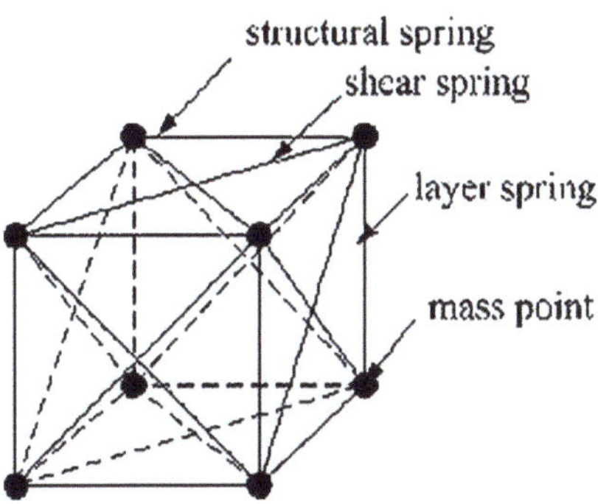

Fig. 1.2 Hexahedral volumetric mass spring model

by structural springs and diagonal shear springs as shown in Fig. 1.2. Structural springs stand for tensile and compressive stress, while shear springs are used for shear stress. Layer springs are added to connect neighboring tissue layers. One soft tissue may consist of several mass spring layers depending on its thickness. For simplicity, many examples shown in this paper doesn't show the structural springs.

1.3 Surgical Cutting on Heterogeneous Models

In this paper, we present the 3D node snapping and topological modification techniques to simulate the surgical cutting operation on heterogeneous deformable models. Hexahedral mass spring structure is maintained without producing extra elements during the topological modification. The main objective of the algorithm is to generate heterogeneous soft tissue layers at the cutting site. For the discussion of the proposed algorithm, some definitions are first given as follows:

Surface cut points: If the cutting tool cuts into a deformable model, it intersects the model surface at the surface cut points (e.g., S_1 and S_2 in Fig. 1.3).

Bottom cut points: Tips of the cutting tool are defined as the bottom cut points (e.g., B_1 and B_2), when the cutting tool cuts into a model.

Cutting lines: Line segments between a pair of a surface cut point and a bottom cut point such as S_1B_1 and S_2B_2 are defined as the cutting lines. Vectors $\overrightarrow{S_iB_i}$ are cutting tool orientations.

Cutting path: The cutting path consists of line segments connecting the surface cut points on the model surface.

Cutting bottom: The cutting bottom consists of line segments connecting the bottom cut points inside the deformable model.

Cutting planes: Movements of the cutting tool are represented by a series of cutting planes. There are two triangles formed by two cutting lines if the two cutting lines are

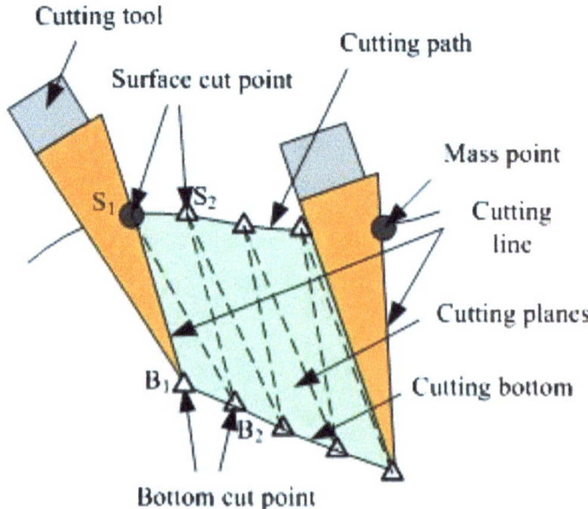

Fig. 1.3 Definitions for volumetric deformable object cutting

not parallel. For example, two planes represented by triangle $S_1S_2B_2$ and $S_1B_1B_2$ exist between cutting lines S_1B_1 and S_2B_2. Since the cutting tool is often manipulated approximately perpendicular to the model surface, the two triangles are approximately parallel. In this paper, the triangle including two surface cut points such as triangle $S_1S_2B_2$ is used to approximate the cutting plane between two cutting lines.

The cutting algorithm is implemented in the following four phases: (1) surface cut snapping, (2) inside cut snapping, (3) cut open generation, (4) internal triangle generation. Details of the algorithm are explained in the following sections.

1.3.1 Surface Cut Node Snapping

In this step, mass points close to the cutting path are snapped along the cutting path on the deformable object surface. In the snapping approach, some springs are aligned along the cutting path, which are used to generate the cut opening. First, surface cut points are sampled when the cutting tool intersects the model surface. The slower the cutting tool moves, the more surface cut points can be recorded. For example, more surface cut points are obtained at slower cutting speed in Fig. 1.4b than those obtained at higher cutting speed in Fig. 1.4a. Intersection points between the cutting path and springs on the model surface are then calculated as shown in Fig. 1.5a. Intersection points are dependent not on the cutting speed but on the cutting path. For each intersection point, the closest mass point is snapped to the intersection point. As a result, some springs are displaced along the cutting path. The snapping and cutting is a progressive process. Before the cutting tool intersects a spring, the last cutting tool position should be updated as the temporary cutting end point. And the cut will be generated and temporarily ends here. This is especially important for visual realism when the cutting tool moves very slowly and the cut should still be updated in the real time.

In this method, only structural springs are used to calculate intersection points due to following two reasons. First, fewer springs are used for intersection testing to reduce the computational burden. More importantly, intersection with diagonal springs may result in degeneracy that all three vertexes of a triangle are snapped into the cutting line. For example, triangles *abc* and *abd* are snapped to the cutting path

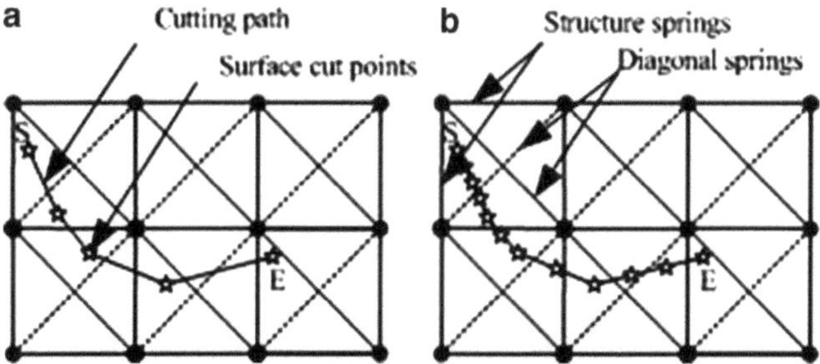

Fig. 1.4 Surface cut points sampling at different cutting speed and intersection points

and both triangles disappear as shown in Fig. 1.5b. Degeneracy can cause the instability of deformable object simulation [13]. On the contrary, such degeneracy can be eliminated by allowing only structural springs for intersection, as shown in Fig. 1.5c.

During the surface cut snapping, a mass point is possibly the closest point to several intersection points. For instance, mass point *A* in Fig. 1.6a is the closest point to both intersection points *B* and *C*. In this case, the midpoint between the first intersection point and the last one is calculated and the mass point is snapped to that midpoint, as point *D* in Fig. 1.6a. However, the first and last cutting points (points *S* and *E*)

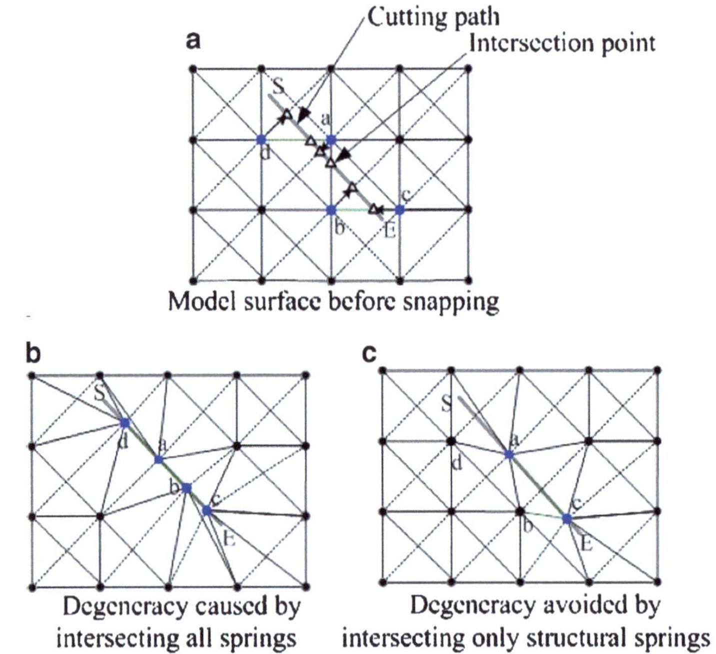

Fig. 1.5 Triangle degeneracy by snapping and the solution

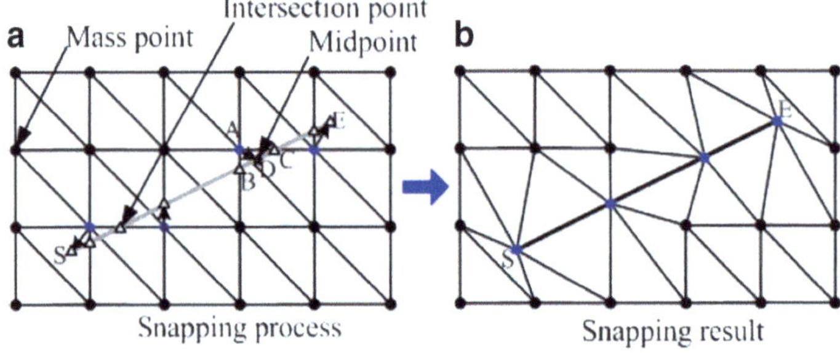

Fig. 1.6 Special surface cut point snapping cases

are treated as special "intersection points." The closest mass points are directly snapped to the first and last cutting points to ensure exact starting and ending locations of the cut, as shown in Fig. 1.6b.

1.3.2 Inside Cut Node Snapping

When cutting into a discrete mass spring model, the tip of the cutting tool may reach a location without mass point or spring definition. This problem can be solved by extending node snapping approach from surface models to volumetric models. The objective is to snap some mass points and associated springs onto the cutting planes (as defined earlier in Fig. 1.3). Figure 1.7 shows an illustrative diagram of inside cut snapping. When the cutting tool moves from S_1B_1 to S_4B_4, surface cut snapping is performed first and mass points S_1, S_2, S_3 and S_4 are snapped along the cutting path, as discussed in Sect. 1.3.1. Then, inside cut snapping is accomplished by following two steps:

1. Mass points are first snapped onto each cutting line, e.g., S_1B_1 and S_4B_4 in Fig. 1.7. Mass points b_1 and b_4 are snapped to the bottom cut points B_1 and B_4 If there are mass points between surface cut points and bottom cut points such as a_1 between S_1 and b_1, these mass points are projected onto the cutting lines as projection points like A_1.
2. It is possible that some mass points may exist between cutting lines such as mass points a_1, a_3, b_2, and b_3 in Fig. 1.7. These mass points and the associated springs are also projected onto the cutting planes, such as $a_{2,\text{projection}}$ and $b_{2,\text{projection}}$ in Fig. 1.8. First, the intersection points between the cutting bottom (e.g., B_1B_4 and the projected springs are calculated. The closest mass points are snapped to the intersection points, such as B_2 and B_3. This can make sure that some springs are aligned along the cutting bottom. The intersection points (e.g., B_1) and their

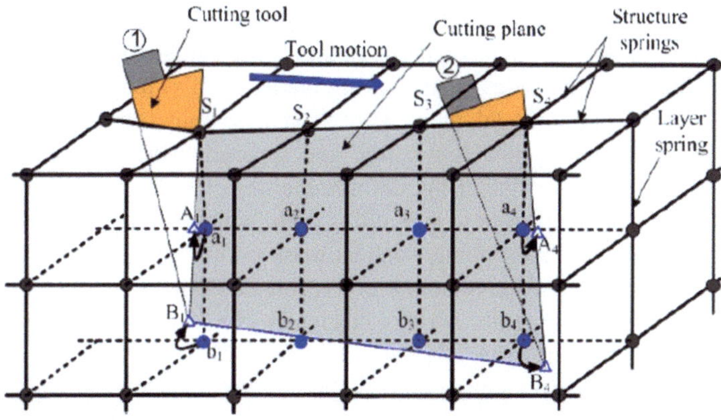

Fig. 1.7 Inside cut point snapping

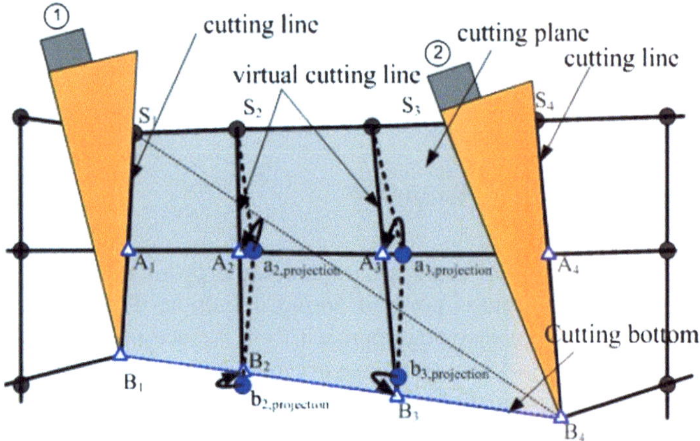

Fig. 1.8 Inside cut snapping onto the cutting plane

corresponding surface snapped points (e.g., S_2) are connected to form the "virtual" cutting lines such as S_2B_2. "Virtual" cutting lines are not the real sampling locations of the cutting tool. At last, other projected mass points on the cutting planes are projected again onto the "virtual" cutting lines. For instance, $a_{2,projection}$ is projected on the "virtual" cutting line S_2B_2 at A_2, as shown in Fig. 1.8.

So far the snapping method only deals with the snapping case in one soft tissue. If more than one tissue layers are involved in one cut, the mass point snapping is quite different and more complex, which will be discussed later in Sect. 1.3.4.

1.3.3 Open Surface Node Modification

After some mass points and springs are snapped on the cutting planes, the cut is generated by dividing the model along the cutting planes. First, snapped mass points, except for the starting and ending points of the cut, are duplicated twice and directly displaced at two sides of the cutting planes such as mass points S_{i1} and S_{i2} for S_i in Fig. 1.9b. In our model, the mass of each mass point is represented by mass density [13]. The mass point duplication and separation don't change the mass density so that the total mass of the deformable object is reserved. The displacement direction is perpendicular to the cutting path, which will be explained shortly. Then original snapped mass points such as mass point S_i are deleted. All springs connected to the deleted mass points are reconnected to their duplicated mass points. As shown in Fig. 1.9b, springs originally connected to mass point S_i are connected to mass point S_{i1} or S_{i2}, depending on which side of the cut they are located at. In particular, both mass point S_{i1} and S_{i2} are connected to the starting cut point S. After some mass points on the model surface are snapped, duplicated and deleted, surface triangles in the vicinity of the cut need to be updated. Triangles are

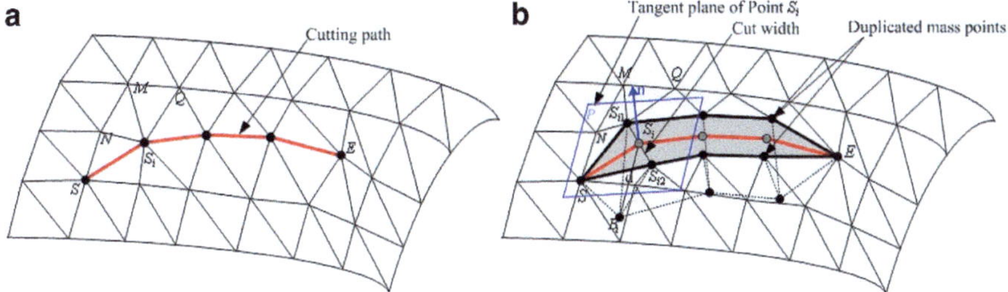

Fig. 1.9 Cut opening generation: (**a**) after surface node snapping; (**b**) after opening the cut

updated according to the reconnected springs near the cut. For example, triangles SS_iN, S_iMN, and S_iQM in Fig. 1.9a are updated to $SS_{i1}N$, S_{i1}, MN, and S_{i1}, QM in Fig. 1.9b, after S_i is replaced by S_{i1} and S_{i2}.

The displacement vector is defined on the tangent plane of the original mass point, such as the tangent plane P of mass point S_i in Fig. 1.9. The tangent plane normal of a mass point is calculated by the average normal of its neighboring triangles by

$$\overrightarrow{n_i} = \frac{1}{m}\sum_{j=0}^{m}\overrightarrow{n_{ij}} \tag{1.1}$$

where m is the neighboring triangle number of mass point i and $\overrightarrow{n_{ij}}$ is the normal of neighboring triangle j. Let w_i be the cut width at mass point i, which can be calculated as

$$w_i = c\frac{d_i - h_i}{1-\alpha_i} \tag{1.2}$$

where α_i is the initial surface tension($0.0 \le \alpha_i \le 1.0$); d_i is the cutting depth; h_i is the depth of the mass point i from the model surface; c is a user-defined scaling parameter to control the cut width. For mass points on the model surface, h_i is zero so that w_i has the maximal value as $w_{s,i}$ in Fig. 1.10. At the cutting bottom, h_i is equal to d_i so that $w_{s,i}$ becomes zero. It remains an open question in Biomechanics to determine the initial skin surface tension [12]. Generally, larger α_i means higher surface tension and wider cut open. On the other hand, the cut width cannot be unlimitedly large, which can be easily observed in the real soft tissue incision. A threshold value w_{max} is applied to limit the width of cut. In general, w_{max} is slightly smaller than the average of spring length. Equation (1.2) is then modified as following:

$$w_i = \min\left(c\frac{d_i - h_i}{1-\alpha_i},\ w_{\max}\right) \tag{1.3}$$

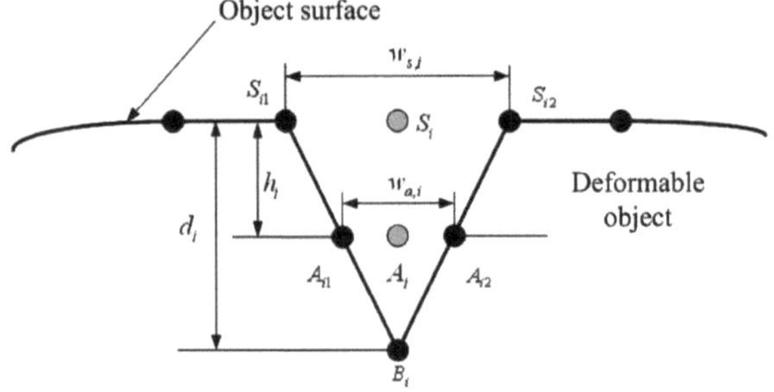

Fig. 1.10 Cross section of the cut opening

The displacement of duplicated mass points can be calculated by following equations:

$$\begin{aligned}\overrightarrow{S_iS_{i1}} &= \frac{w_i}{2}\frac{\vec{n}_i \times \overrightarrow{SS_i}}{\left|\vec{n}_i \times \overrightarrow{SS_i}\right|}\\ \overrightarrow{S_iS_{i2}} &= -\frac{w_i}{2}\frac{\vec{n}_i \times \overrightarrow{SS_i}}{\left|\vec{n}_i \times \overrightarrow{SS_i}\right|}\end{aligned} \tag{1.4}$$

And new positions of duplicated mass points can be calculated by (1.5):

$$\begin{aligned}S_{i1} &= S_i + \min\left(c\frac{d_i - h_i}{(1- \;_i)},\ w_{\max}\right)\frac{\vec{n}_i \times \overrightarrow{SS_i}}{\left|\vec{n}_i \times \overrightarrow{SS_i}\right|}\\ S_{i2} &= S_i - \min\left(c\frac{d_i - h_i}{(1- \;_i)},\ w_{\max}\right)\frac{\vec{n}_i \times \overrightarrow{SS_i}}{\left|\vec{n}_i \times \overrightarrow{SS_i}\right|}\end{aligned} \tag{1.5}$$

Although this cut opening method does not follow the physical law, the generated cut is unconditionally smooth. The cut result is visually smoother than the traditional zig-zag cut edge shown in earlier work [4, 8, 9]. Furthermore, this method is more efficient in computation than the physically based numerical integration methods where the cut is generated by the spring forces of disconnected springs.

1.3.4 Internal Triangle Generation

After the cut is opened by the method in Sect. 1.3.3, internal triangles at both sides of the cut are generated to form watertight surfaces for the deformable model. Different from groove generation algorithms of previous work that were applied to homogeneous surface models, the proposed internal triangle generation method for heterogeneous volumetric deformable models features representing different soft tissues layers explicitly at the cutting site. Figure 1.11 shows an illustrative diagram of internal triangle generation when the cutting tool cuts from position 1 to position 2 at one tool movement. The model is separated along the cutting plane (A–A plane) to show the inside details in Fig. 1.12. The internal triangle generation method has following two cases as shown in Fig. 1.12.

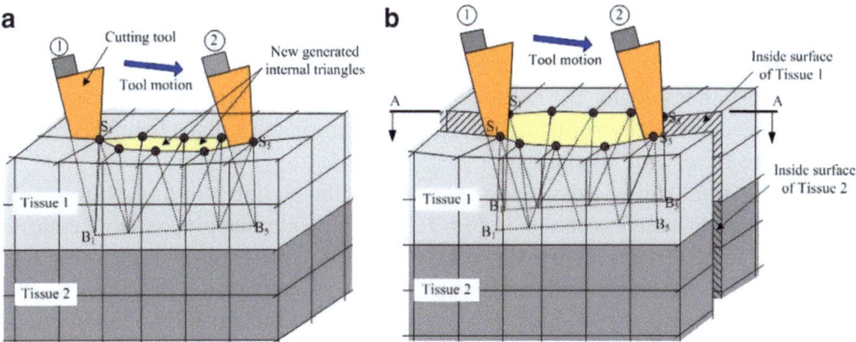

Fig. 1.11 Cutting on heterogeneous model

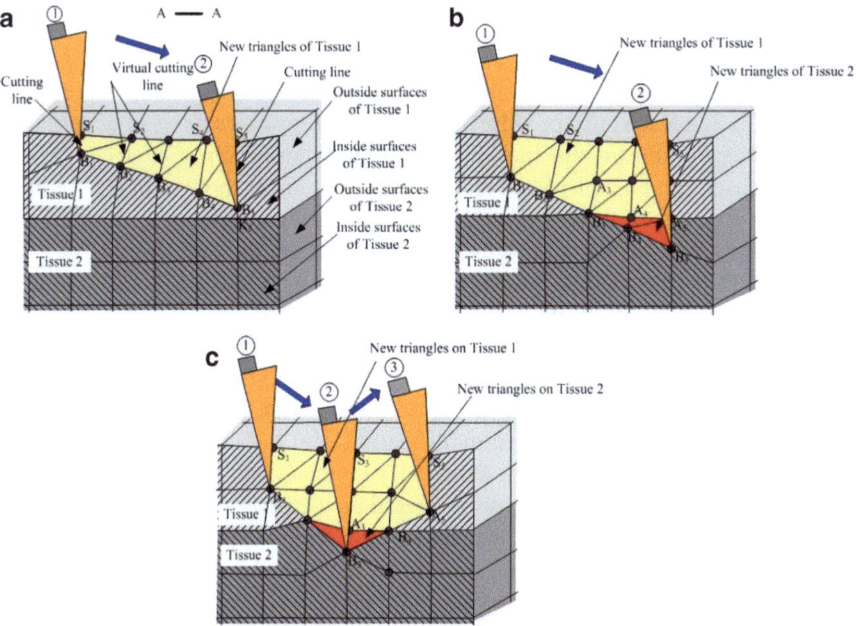

Fig. 1.12 Cutting cases on the heterogeneous model

Case 1: Only one tissue is cut when the cutting tool moves from S_1B_1 to S_5B_5 as shown in Fig. 1.12a. The cut is not deep enough to reach the second tissue layer (tissue 2). For heterogeneous models, the node snapping is different near the interface between two soft tissues. It is possible that the cutting tool is very close to the interface surface but hasn't reached the interface yet. The closest mass point on the interface is not snapped to the bottom cut point. In Fig. 1.12a, although mass point K_5 is closer to B_5 than other mass points, it is not snapped to B_5. Otherwise, it would indicate that the cutting tool has reached the interface surface. After the inside cut snapping, the cut is opened as discussed in Sect. 1.3.3. New internal triangles need to be generated at both sides of the cut. In Case 1, internal triangles are generated between cutting lines and virtual cutting lines as shown in Fig. 1.12a. For example, new triangles are generated by the vertex sequence of $S_1B_1S_2$ and $S_2B_1B_2$. Only one side of the cut is shown in Fig. 1.12a and triangles are also generated similarly at the other side of the cut.

Case 2: More than one tissue layers are cut as the cutting tool moves from S_1B_1 to S_5B_5 at one time step as shown in Fig. 1.12b. The node snapping is more complicated near the interface of two tissue layers. (1) When the tip of the cutting tool reachesB_3, it hasn't cut into tissue 2. After that, the cutting tool cuts into tissue 2 and its tip reaches B_4 and B_5 later. In this situation, B_3 is snapped onto the interface surface. (2) When the cutting tool cuts into another soft tissue and no matter how deep it is, a mass point of that soft tissue should be snapped to the tip of the cutting tool to represent the cutting depth. For example, even if the mass point A_4 is closer to B_4, a mass point of tissue 2 is snapped to $B_4$to ensure the correct cutting depth information.

After the mass point snapping, new triangles are generated at two sides of the cut, similar to the method in the first case. Different sets of triangles are separated by the interface surface. More complex cutting cases exist in accordance with various cutting depths such as the example shown in Fig. 1.12c. These complex cases can be considered as the combinations of the two cases we have discussed. The material properties are assigned to these new generated internal triangles according to their vertices' properties. We argue that vertices of a triangle, except for vertices on the model surface or an interface surface, belong to only one type of tissue in our deformable models. For example in Fig. 1.12b, triangle $S_1B_1S_2$ belongs to tissue 1 because B_1 is mass point of tissue 1 and S_1 and S_2 are on the model surface. Triangle $B_3B_4A_4$ belongs to tissue 2, since B_3 and B_4 are mass points of tissue 2 and A_4 is on the interface surface. Although the node snapping approach for topological modification does not increase the number of mesh elements, degeneracy such as smaller element size or badly shaped elements may still occur after the node snapping. Degeneracy can cause system instability of downstream numerical integration for the deformation simulation. A quasi-static algorithm has been developed to implement mesh refinement in our previous work [16]. Mass points are iteratively relocated based on their spring forces.

1.4 Implementation and Results

The proposed cutting techniques are implemented on a 2.4 GHz dual-CPU workstation with 2 GB memory and a high-end graphics card. Microsoft® Visual C++6.0 and Open GL® library are the fundamental development tools. The system is integrated with a 6-DOF (degree of freedom) Phantom® desktop, which can provide 3-DOF force feedback. The integration of haptic devices into virtual systems has been discussed in our previous work [17, 18]. Figure 1.13a shows the system setup for cutting simulation. During the cutting process, topology of the deformable model is modified. In the mean time, the model is deformed by the cutting tool. Numerical integration of the simulation system is based on constrained local static integration method in our previous work [13]. Each mass point is iteratively moved under its spring force but constrained within the bounding box defined by its neighboring mass points. Figure 1.13b shows an example of cutting a heterogeneous deformable model consisting of a total of 4,152 mass points by using a virtual cutting tool.

Figure 1.14 shows a cut opening result on the model surface. New springs are generated and connected to their neighboring mass points, as shown in the lower right figure. Internal triangles are generated to form a watertight surface as shown in the lower left figure. The cut is smooth as shown in Fig. 1.14.

Figure 1.15 shows the heterogeneity of the deformable model at the cutting site. Fat tissues and muscles are represented by different colors. Thickness of each tissue layer depends on geometry of the deformable model. The cutting result on heterogeneous model is able to provide users with most important visual feedback such as color or texture. On the other hand, heterogeneous material properties like stiffness on each layer also gives the user appropriate tactile cue. With realistic visual and haptic feedback, users can easily control their cutting behaviors during the cutting.

a

b

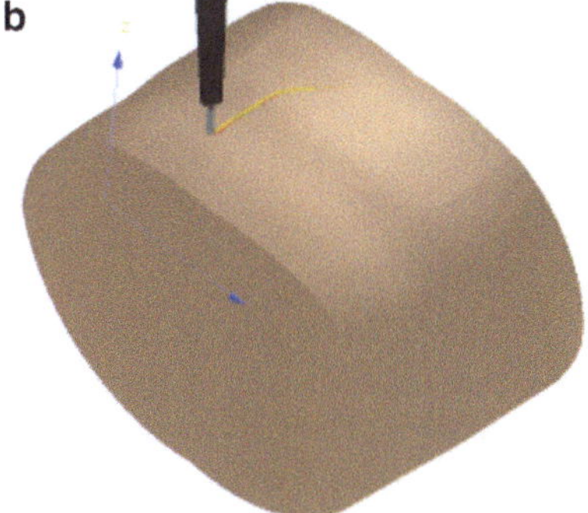

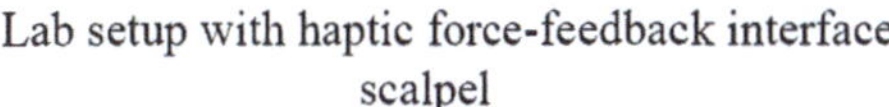

Lab setup with haptic force-feedback interface

Cutting simulation using a surgical scalpel

Fig. 1.13 Cutting simulation using a Phantom haptic device

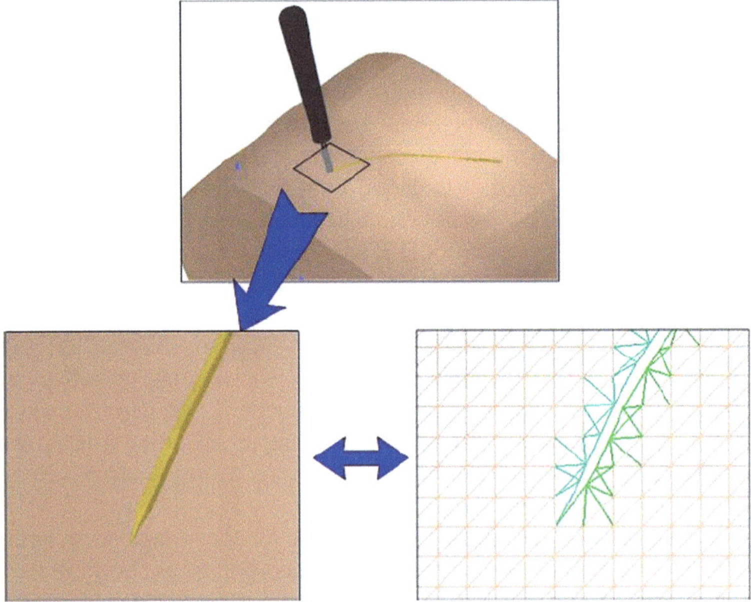

Fig. 1.14 Arbitrary cutting result and new generated springs at cutting site

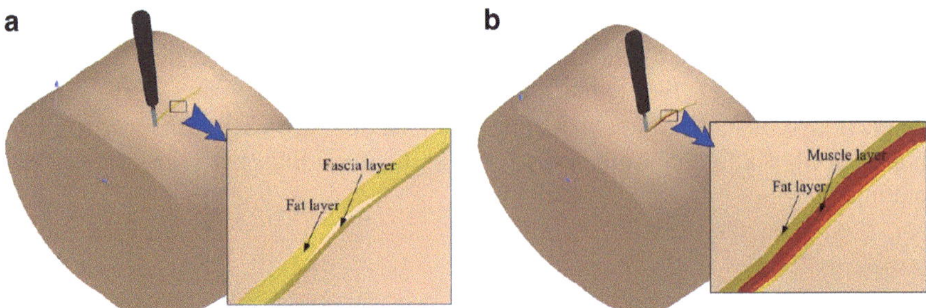

Fig. 1.15 Multiple layers at different cutting sites: (**a**) cutting into right rectus, (**b**) cutting into linea alba

Figure 1.16 shows the process of cutting soft tissues on a heterogeneous leg model to expose the bone (tibia). Skin and part of fascia are cut open first, as shown in Fig. 1.16a. Then the model is cut deeper inside until the tibia is reached. Because bones are considered as rigid objects, they are not cut open by a general cutting tool like a scalpel. Users can see the intact bone surfaces and perform further surgery on bones.

Figure 1.17 shows an example of continuous cuts on a human leg model. After the model is cut first as shown in Fig. 1.17b, new cut can be done on the existing cut either to deepen or to lengthen the cut. Figure 1.17c shows an example

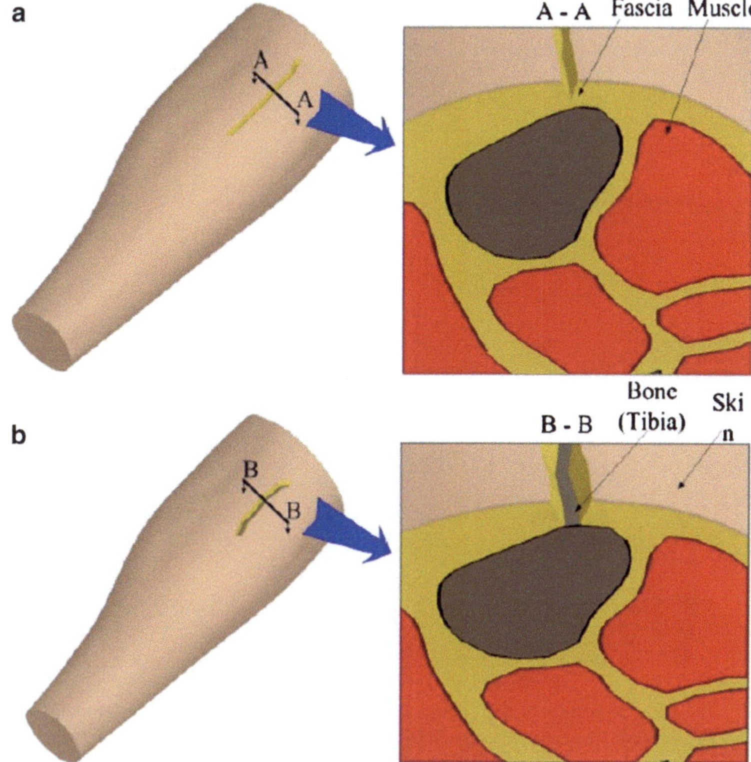

Fig. 1.16 Cutting results on different cutting depths

of making the old cut deeper. As a result, the muscle layer is exposed by the second cut. The cut is further deepened as shown in Fig. 1.17d. This provides users the training scenario of cutting the soft tissues layer by layer as the surgeons do in real surgery.

1.5 Conclusion

In this paper, cutting techniques are proposed for cutting simulation on volumetric heterogeneous deformable models. 3D node snapping and topology modification approaches are presented to generate the smooth cut. The cut can present heterogeneous interior structures and material properties of deformable models. The proposed cutting techniques can enhance the fidelity and realism of surgical simulation. They can also be used to other virtual simulation systems when topological modification is involved.

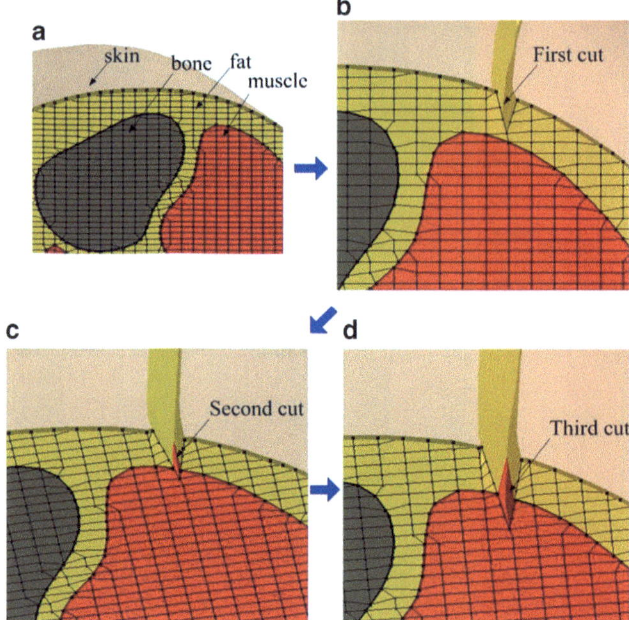

Fig. 1.17 The detailed results of continuous cuts on a leg model. (**a**) original model; (**b**) cutting result after the first cut; (**c**) cutting result after the second cut; (**d**) cutting result after the third cut

Acknowledgment This work was partially supported by the National Science Foundation (NSF) Grant (CMMI-0553310 and CMMI-0800811) to North Carolina State University and University of North Carolina at Chapel Hill. Their support is greatly appreciated.

References

1. Mosegaard J (2004) Realtime cardiac surgery simulation. PhD Dissertation, Aarhus University Hospital, Denmark
2. Fuller JR (1986) Surgical technology principles and practice, 2nd edn. W. B. Saunders Company, Philadelphia, PA
3. Bielser D, Maiwald VA, Gross MH (1999) Interactive cuts through 3-dimensional soft tissue. Comput Graph Forum (Eurographics 99 Proc) 18(3):31–38
4. Bruyns CD, Senger S, Menon A, Montgometry K, Wildermuth S, Boyle RA (2002) Survey of interactive mesh-cutting techniques and a new method for implementation generalized interactive mesh cutting using virtual tools. J Visualiz Comput Anim 13:21–42
5. Bro-nielsen M, Helfrick D, Glass B, Zeng X, Connacher H (1998) VR simulation of abdominal trauma surgery. In: Proceedings of medicine meets virtual reality 6, San Diego, California, 28–31 Jan 1998, pp 117–123

6. Lim Y–J, De S (2004) On the use of meshfree methods and a geometry based surgical cutting algorithm in multimodal medical simulations. In: Proceedings of the 12th international symposium on haptic interfaces for virtual environment and teleoperator systems (HAPTICS'04), pp 295–301
7. Zhang H, Payandeh S, Dill J (2004) On cutting and dissection of virtual deformable objects. In: Proceedings of the 2004 IEEE international conference on robotics & automation, Apr 2004, pp 3908–3913
8. Delingette H, Cotin S, Ayache N (1999) A hybrid elastic model allowing real-time cutting, deformations and force-feedback for surgery training and simulation. In: Proceeding computer animation, 70–81
9. Nienhuys H-W, van der Stappen AF (2001) A surgery simulation supporting cuts and finite element deformation. In: Proceedings medical image computing and computer assisted intervention (MICCAI), 145–152
10. Lin S, Lee Y-S, Narayan R, Shin H (2007) Bio-tissues modeling and interface development for bio-manufacturing and medical applications. In: Proceedings of the 2007 industrial engineering research (IERC) conference, Nashville, TN, 19–23 May 2007, pp 281–286
11. Lin S, Lee Y-S, Narayan R (2007) Heterogeneous soft material modeling and virtual prototyping with 5-DOF haptic force feedback for product development. In: Proceedings of the 3rd international conference on advanced research in virtual and rapid prototyping, (VRAP 2007), Leiria, Portugal, 24–28 Sept, 2007, pp 187–193
12. Fung YC (1993) Biomechanics: mechanical properties of living tissues, 2nd edn. Springer-Verlag, New York
13. Lin S, Lee Y-S, Narayan R (2007) Snapping algorithm and heterogeneous bio-tissues modeling for medical surgical simulation and product prototyping. Virtual Phys Prototyp 2(2):89–101
14. Meier U, Lopez O, Monserrat C, Juan MC, Alcaniz M (2005) Real-time deformable models for surgery simulation: a survey. Comput Meth Prog Biomed 77:183–197
15. Mollemans W, Schutyser F, Cleynenbreugel JV, Suetens P (2003) Tetrahedral mass spring model for fast soft tissue deformation. In: Proceedings of surgery simulation and soft tissue modeling: international symposium IS4TM, pp 145–154
16. Lin S, Narayan R, Lee Y-S (2008) Heterogeneous deformable modeling and topology modification for surgical cutting simulation with haptic interfaces. Computer-Aid Design Appl 5(6):877–888
17. Lin S, Narayan R, Lee Y-S (2007) Collaborative haptic interfaces and distributed control for product development and virtual prototyping. In: Proceedings of 2007 international manufacturing science & engineering conference, Atlanta, GA, USA, 15–18 Oct 2007, Paper number: MSEC2007-31214
18. Lin S, Lee Y-S, Narayan R (2008) Heterogeneous material modeling and virtual prototyping with 5-DOF haptic force feedback for product development. Int J Mechatron Manufactur Syst 1(1):43–67

Chapter 2
Heterogeneous Deformable Modeling of Bio-Tissues and Haptic Force Rendering for Bio-Object Modeling

Shiyong Lin, Yuan-Shin Lee, and Roger J. Narayan

Abstract This paper presents a novel technique for modeling soft biological tissues as well as the development of an innovative interface for bio-manufacturing and medical applications. Heterogeneous deformable models may be used to represent the actual internal structures of deformable biological objects, which possess multiple components and nonuniform material properties. Both heterogeneous deformable object modeling and accurate haptic rendering can greatly enhance the realism and fidelity of virtual reality environments. In this paper, a tri-ray node snapping algorithm is proposed to generate a volumetric heterogeneous deformable model from a set of object interface surfaces between different materials. A constrained local static integration method is presented for simulating deformation and accurate force feedback based on the material properties of a heterogeneous structure. Biological soft tissue modeling is used as an example to demonstrate the proposed techniques. By integrating the heterogeneous deformable model into a virtual environment, users can both observe different materials inside a deformable object as well as interact with it by touching the deformable object using a haptic device. The presented techniques can be used for surgical simulation, bio-product design, bio-manufacturing, and medical applications.

2.1 Introduction

The techniques of modeling of biological tissues and deformable objects for bio-manufacturing and medical applications have not been successfully dealt with using conventional processes. Many biological tissues possess multiple components

S. Lin and Y-S. Lee (✉)
Edward P. Fitts Department of Industrial and Systems Engineering, North Carolina State University, Raleigh, NC 27695, USA
e-mail: yslee@ncsu.edu

R.J. Narayan
Department of Biomedical Engineering, University of North Carolina, Chapel Hill, NC 27599, USA

R. Narayan et al. (eds.), *Printed Biomaterials*, Biological and Medical Physics, Biomedical Engineering, DOI 10.1007/978-1-4419-1395-1_2,

and complex geometric information. Figure 2.1 shows the boundary surfaces and interface surfaces of the human anterior abdominal wall. As seen in this figure, adipose tissue, skin, and muscle tissues form a multilayered structure in this region of the body. It is difficult to model these structures using conventional approaches. Physically based deformable models are also of growing interest in virtual reality (VR) applications, including surgery, entertainment, design, and manufacturing, since most deformable objects consist of heterogeneous materials with complex internal structures [1–4]. Deformable models must incorporate appropriate internal structures and material properties for situations in which high-fidelity virtual environments are required [5].

Heterogeneous deformable model representation is critical in modeling internal structures of deformable biological objects that possess multiple components with different material properties [6,7]. In previous studies, nonlinear elasticity of deformable objects has been modeled using nonlinear strain tensors or nonlinear spring coefficients [8,9]. Bourguignon [10] and Picinbono [11] reported the modeling of anisotropic behaviors of deformable objects. However, these approaches are only capable of dealing with deformable objects that consist of single material type.

Haptic interfaces have attracted a lot of research interests in the recent years. Haptic interface is an electromechanical device that can provide force feedback to users [12,13]. Haptic force-feedback interface has been used for conceptual design, collaborative design, virtual prototyping, and sculpting in VR environments [14–16]. Heterogeneous deformable object models combined with accurate haptic rendering techniques based on these models provide great potential to enhance interactions with deformable objects. For example, in medicine and surgery it is often necessary to differentiate healthy soft tissues from unhealthy ones [5,17]. However, most current practices assume homogeneous material properties and ignore internal structure variations, which greatly limited their use [18]. In VR environments, new deformable models often need to be generated efficiently. In our earlier work presented in [4,19], haptic-based force rendering techniques have been

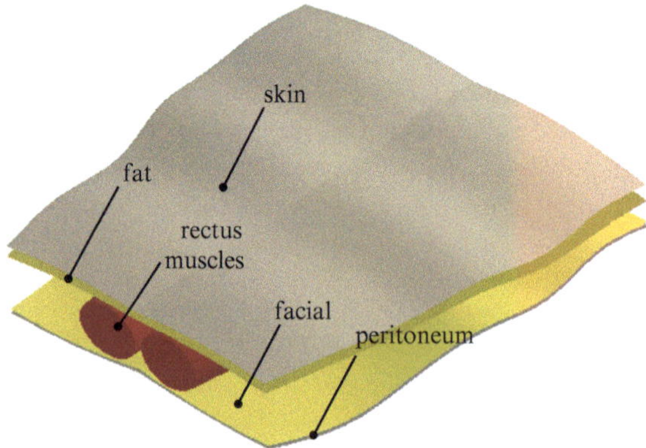

Fig. 2.1 Bio-tissue example and boundary surfaces of anterior abdominal wall

developed for medical surgical applications. For example, patient-specific models of soft tissues may be required for the presurgical rehearsal and postsurgical analysis. However, conventional virtual simulation systems lack appropriate flexibility for creating deformable models, especially heterogeneous deformable models.

In this paper, detailed techniques and modeling algorithms are proposed to generate heterogeneous deformable models for bio-manufacturing and medical applications. A constrained local static integration method is proposed for rapid and robust deformation simulation and a force accumulation model is presented to calculate spring forces for accurate haptic rendering. Parts of human body such as anterior abdominal wall, leg, and thigh are modeled using the proposed techniques for demonstration. Other heterogeneous deformable objects can also be modeled using these techniques.

The remainder of this paper is organized as follows. Section 2.2 provides a review of heterogeneous modeling of deformable objects. Section 2.3 presents the detailed node snapping algorithm for modeling deformable heterogeneous objects. Section 2.4 discusses the system modeling of spring force rendering and the detailed techniques of solving the deformation equilibrium. Section 2.5 presents the computer implementation and practical examples of the proposed methods, followed by the concluding remarks in Sect. 2.6.

2.2 Modeling of Deformable Biological Soft Tissues

Biological soft tissues are viscoelastic, anisotropic, and heterogeneous deformable objects [20]. Deformable model generation is a process of discretizing volumetric objects into mass points and springs in a Mass Spring Model or small elements in a Finite Element Model. Most simulation systems use the mesh generation approaches of Finite Element Analysis (FEA) to generate small elements such as tetrahedra or hexahedra for a Finite Element Model [21]. Many Mass Spring models can also be generated from these tetrahedra or hexahedra by taking their vertices as mass points and edges as springs [22]. However, homogeneous deformable models are generated using these approaches in many cases.

A heterogeneous deformable object may include several materials, as shown in Fig. 2.2. For example, a human leg consists of skin, adipose tissue, muscles, bone, tendons, and other tissues. To generate such a heterogeneous model, the geometry of each material inside the object is required, such as its interface surfaces. Most interface surfaces between materials can be acquired from computed tomography, magnetic resonance imaging, and other techniques, although it remains as a challenge to differentiate some similar tissues.

In some applications like Computer Aided Design (CAD), smooth variation between different materials is traditionally applied by interpolation [23]. For heterogeneous deformable modeling of biological soft tissues, a meaningful interface layer usually exists between two different materials, as shown in Fig. 2.2. For instance, muscles are often enclosed by the muscle sheath and bones are enclosed

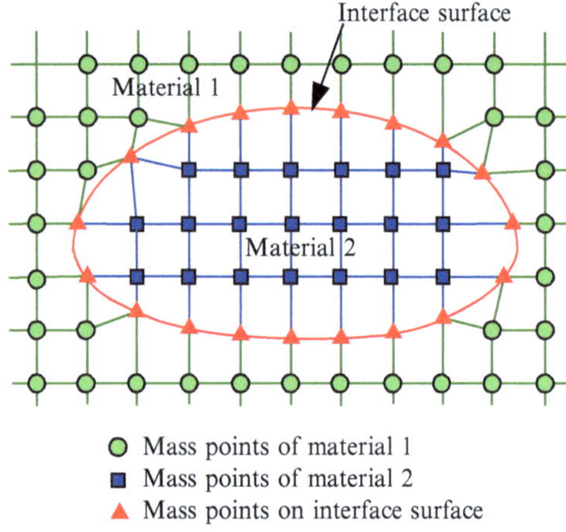

Fig. 2.2 Modeling of heterogeneous deformable bio-objects with interface surface between tissues

by the periosteum. These interface surfaces provide appropriate visual clue to differentiate soft tissues. They can also be used for specifying different material properties such as mass, texture, and stiffness.

Heterogeneous modeling and simulation require more computational power than homogeneous modeling since more topological constraints and material properties are involved. For interactive VR applications (e.g., surgical simulation), real-time computation has the highest priority. As a result, algorithms used in heterogeneous modeling must be as efficient as possible. In addition, the balance between the level of details in heterogeneous modeling and computational efficiency needs to be carefully taken into account.

2.3 Node Snapping Algorithm for Constructing Heterogeneous Models

In this paper, we present a technique of constructing heterogeneous deformable models. The key point of heterogeneous model generation is to retain a single mass spring layer for the outmost surface of a whole object and also individual mass spring layers for each internal interface surface. For the remainder of the deformable model, a uniform mass spring network is created to connect these interface mass spring layers. For a better illustration of the idea of heterogeneous deformable model generation, two definitions are provided below:

Definition 1: Mass points of the same material are defined as homogeneous mass points; otherwise they are heterogeneous mass points.

Definition 2: Springs connecting homogeneous mass points are defined as homogeneous springs; springs connecting heterogeneous mass points are defined as heterogeneous springs.

In this paper, the heterogeneous deformable model generation algorithm is implemented in the following sequence: (1) mass point generation by a tri-ray node snapping algorithm, (2) springs connection and interface marching technique, and (3) physical parameter specification.

From the boundary and interface surfaces, mass points are generated by a tri-ray node snapping algorithm similar to the Ray Casting algorithm in Computer Graphics. On the basis of the concept of the tri-dexel volumetric model developed in our earlier work [12], parallel rays are cast respectively three orthogonal directions to intersect the boundary surface and interface surfaces, to accurately preserve the shapes of these surfaces. Figure 2.3 shows the procedure of using tri-ray tracing

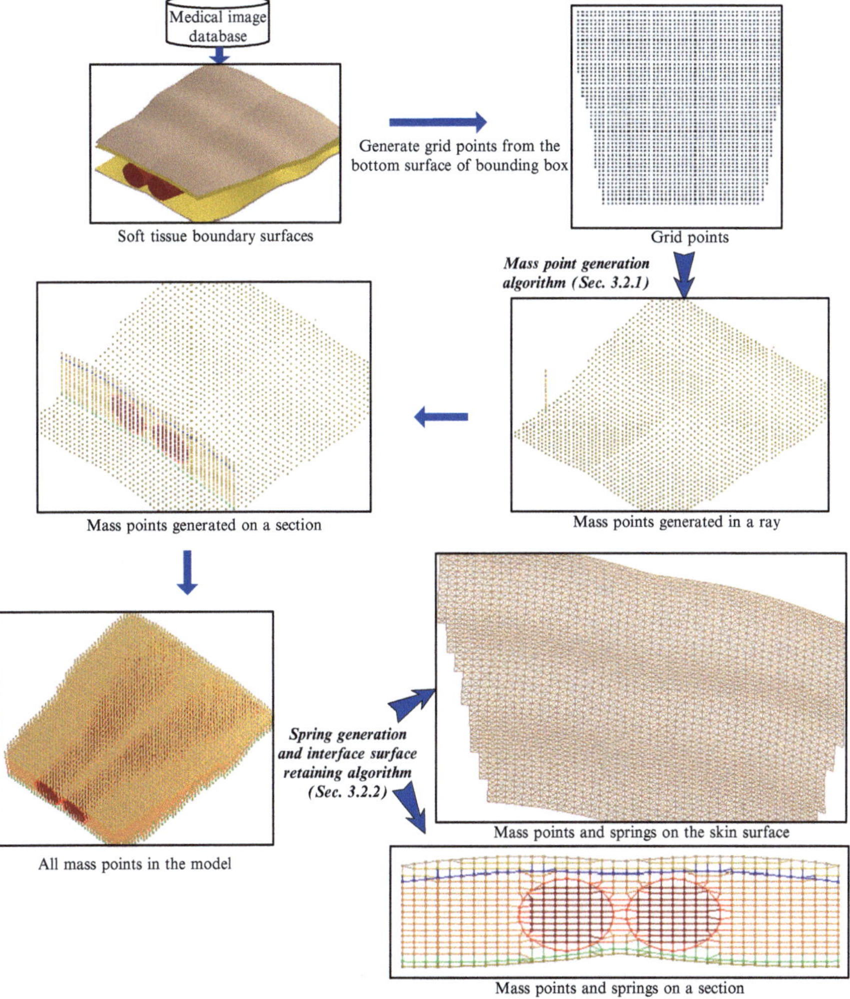

Fig. 2.3 Procedure of constructing heterogeneous volumetric models

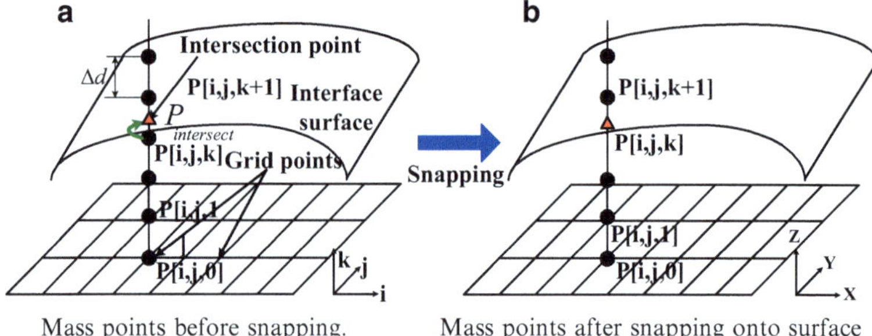

Fig. 2.4 Mass points snapped to intersection surface points, (**a**) Mass points before snapping, (**b**) Mass points after snapping onto surface

technique developed in our earlier work in [5] to construct volumetric heterogeneous models and the technique is briefly explained as follows.

First, rays in one direction are used to generate all mass points. Then rays in the other two directions are used to adjust the positions of mass points around interface surfaces, without generating new mass points. As shown in Fig. 2.4, uniform rays are cast from grid points $P[i, j, 0]$ on the two-dimensional X–Y plane to intersect with all the accessible boundary surfaces. Mass points are evenly generated at the interval Δd along the ray as shown below:

$$P[i,j,k+1] = P[i,j,k] + \Delta d \cdot \vec{r} \tag{2.1}$$

where $\vec{r}$ is a unit vector in the ray direction; Δd is the grid distance between adjacent mass points. Δd is defined by the maximum of d_{min} and $d_{\mathrm{threshold}}$ as follows:

$$\Delta d = max\{d_{threshold}, d_{min}\} \tag{2.2}$$

In (2.2), d_{min} is the minimal distance between two intersection points of different interface surfaces in the model, and $d_{threshold}$ is a preset value to control the mass point density in case d_{min} is too small. As shown in Fig. 2.4a, when a ray intersects an interface surface at the point $P_{intersect}$, the closest mass point to $P_{intersect}$ along the ray is $P[i,j,k]$ if the following condition is true:

$$\left|P_{intersect} - P[i,j,k]\right| \leq \frac{\Delta d}{2} \tag{2.3}$$

Using (2.3), a mass point can be snapped into its closest intersection point on an interface surface. As shown in Fig. 2.4b, $P[i, j, k]$ is the closest mass point to $P_{intersect}$ and it is snapped to $P_{intersect}$ as a result. Figure 2.5 shows the flowchart of the detailed procedures for generating mass points in heterogeneous volumetric models.

Using (2.1)–(2.3), mass points generated by the tri-ray procedure can be refined into the intersection points located on the surfaces. Mass points are gradually generated along the ray until the ray reaches the outmost boundary surface. This

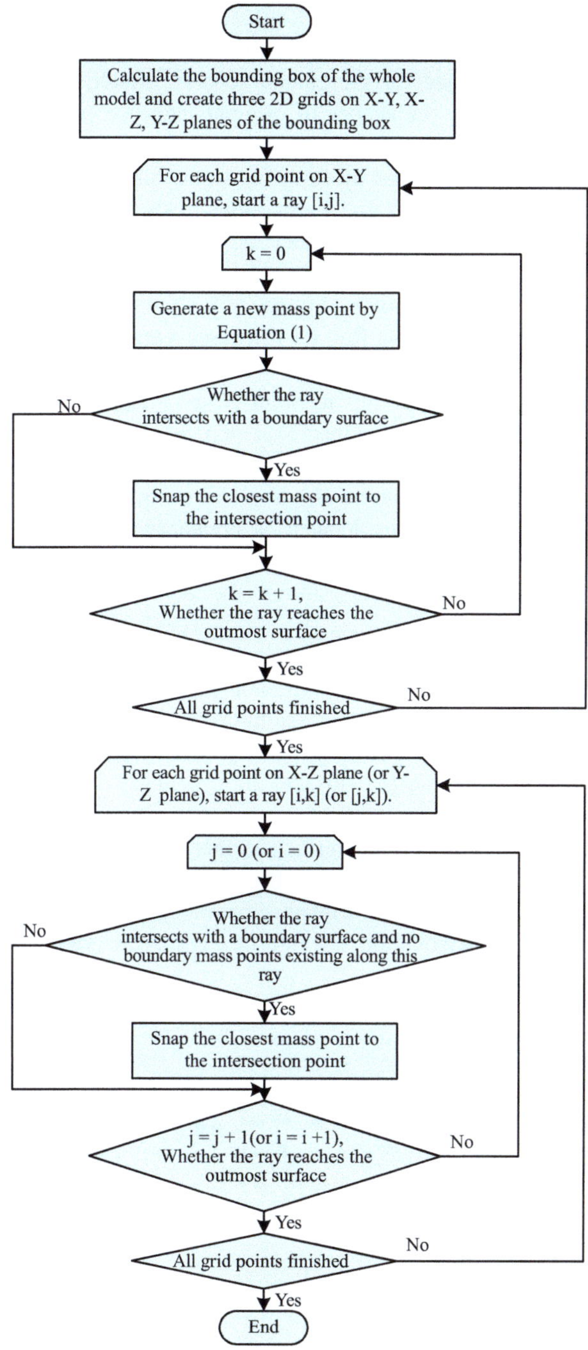

Fig. 2.5 Flowchart of generating the volumetric heterogeneous models

process is repeated until all rays in the *Z* direction are processed. After this step, newly generated mass points are shown in the *X–Z* plane of Fig. 2.6a. However, mass points on the interface surface cannot separate Material 1 and Material 2. They also cannot preserve the shapes of original interface surfaces, as shown in Fig. 2.6a.

These problems may be solved by adjusting some mass points around the interface surfaces using rays in the *X* and *Y* directions, respectively. For example, rays along the *X* direction are cast to intersect the interface surfaces. Existing mass points closest to the intersection points are snapped to these intersection points, without generating new mass points. Figure 2.6b shows an example of adjusting mass points along the *X* direction. Ray $k+1$ intersects the interface surface at point M and point N. Mass point $P[i-1, k+1]$, which is closest to point M, and mass point $P[i+3, k+1]$, which is closest to point N, are snapped to points M and N, respectively. After the mass point adjustment step, a smooth boundary interface is formed to separate the two materials (Fig. 2.6c). Figure 2.7 gives an example of mass points on one section of human thigh model using the tri-ray node snapping algorithm. Larger mass points are located on boundary or interface surfaces. The example shows that the node snapping algorithm can preserve boundary and interface shapes in deformable objects during the construction of heterogeneous deformable models.

Most mass points by the tri-ray node snapping algorithm are uniformly distributed inside the deformable model if they are not snapped onto any interface surface. These mass points maintain a regular neighboring topology so that they can be directly connected to their neighboring mass points by springs. On the contrary, mass points snapped onto interface surfaces are irregularly distributed due to snapping. Mass points snapped on an interface surface are supposed to form a single mass spring layer to represent this interface surface. However, the direct neighboring mass point connection method is not able to achieve this.

Figure 2.8a shows an example of interface points generated by the tri-ray node snapping algorithm. Notice that the new interface points are not located at the exact

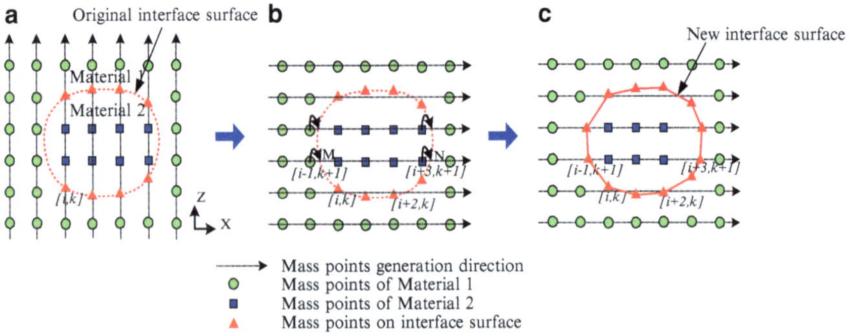

Fig. 2.6 Mass point generation with different bio-tissue materials. (**a**) Generate mass points in Z direction, (**b**) Adjust mass points in X direction, (**c**) Final mass points

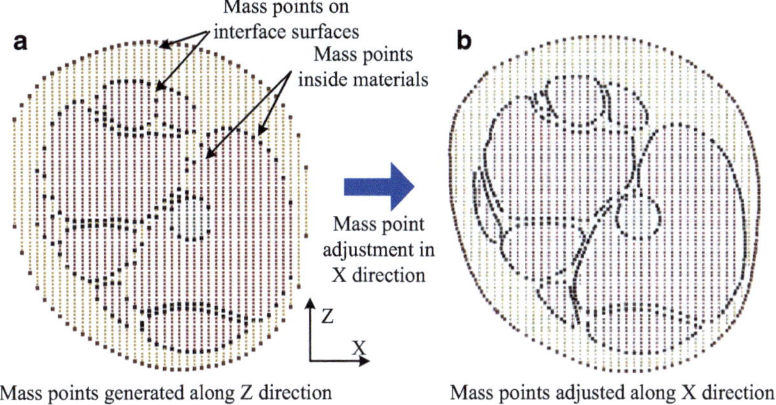

Fig. 2.7 Node snapping algorithm and mass-spring point construction of human thigh example

grid nodes. In this paper, an interface marching technique has been developed to construct the interface surface combined with the node snapping procedure. Using the concept of the marching cube algorithm developed in our earlier work presented in [13], the interface surface is connected through marching along the three-dimensional surface contour. In Fig. 2.8b, the original grid nodes generated by the tri-ray method are snapped onto the interface surface. When marching along the interface contours ceases, the interface contours need to be connected and merged. As shown in Fig. 2.8c, a continuous interface surface has been successfully constructed. Details of the Interface Marching Algorithm are shown in Fig. 2.9.

The last step of heterogeneous deformable model generation is to assign material properties to mass points and springs. In our heterogeneous deformable models, different materials are separated by the interface surfaces. By selecting any mass point, all of its homogeneous mass points can be chosen simultaneously and material properties (e.g., mass) can be specified efficiently, which is an advantage of our heterogeneous model.

Similarly, we can assign physical properties (e.g., elastic modulus) to springs in this model. If a spring is a homogeneous spring that connects two homogeneous mass points, it is either on an interface surface or within a material. Material properties of the interface surface or the material are assigned to the spring. If a spring is a heterogeneous spring, the mass point at one end of the spring must be on an interface surface and the mass point at the other end of the spring belongs to another material. The spring is considered to have material properties of that material. All springs exhibiting the same physical properties can be found starting from any of these springs, since they are always connected with each other. Therefore, their spring properties can be assigned in a batch.

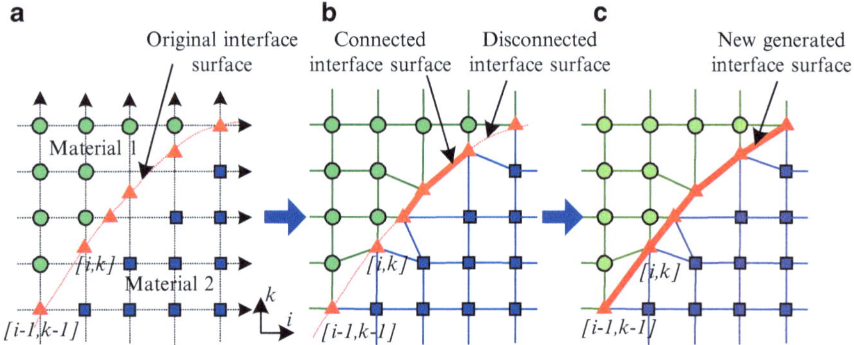

a Mass points generated by the tri-ray snapping algorithm. **b** Spring connected by direct neighboring connection method. **c** Spring generation on the interface surface by homogeneous mass point search algorithm

Fig. 2.8 Technique for Interface Surface Marching. (**a**) Mass points generated by the tri-ray snapping algorithm. (**b**) Spring connected by direct neighboring connection method. (**c**) Spring generation on the interface surface by homogeneous mass point search algorithm

2.4 System Modeling of Spring Force Rendering and Solving Deformation Equilibrium

Local deformation of soft tissues is often assumed in surgical simulators to reduce the computational burden of deformable object simulation [24]. In this paper, local deformation is controlled by a preset small force value. Mass points are considered to be active only if their total spring forces are larger than the threshold force. Only active mass points are used in simulation.

In the conventional Mass Spring Model, the spring force of a mass point i is often calculated by total spring forces from its neighboring springs as follows:

$$f_i = \sum_{t=0}^{n} \frac{k_{it} \Delta l_{it} \vec{d}_{it}}{\left| \vec{d}_{it} \right|} \tag{2.4}$$

where k_{it} is the spring coefficient between mass points i and t; Δl_{it} is the spring length deviation from its original length; $\vec{d}_{it}$ is the direction vector pointing from mass point i to its neighboring mass point t.

During the simulation, it is possible that a spring will become over-compressed or stretched to its extreme length as spring ij in Fig. 2.10. In this case, (2.4) becomes inaccurate because the spring force will not change even if mass point i (Fig. 2.10) is further pushed down. In the extreme compression or stretching case, we assume that a spring will become a rigid object without break or fracture. A force accumulation method is presented to accumulate the spring force acting on the rigid object. For example, spring ij in Fig. 2.10 is extremely compressed and is considered as a rigid object. The spring force on one end mass point (e.g., mass point j) is accumulated to the other end mass point (e.g., mass point i) so that the total spring force at mass point i is calculated as follows:

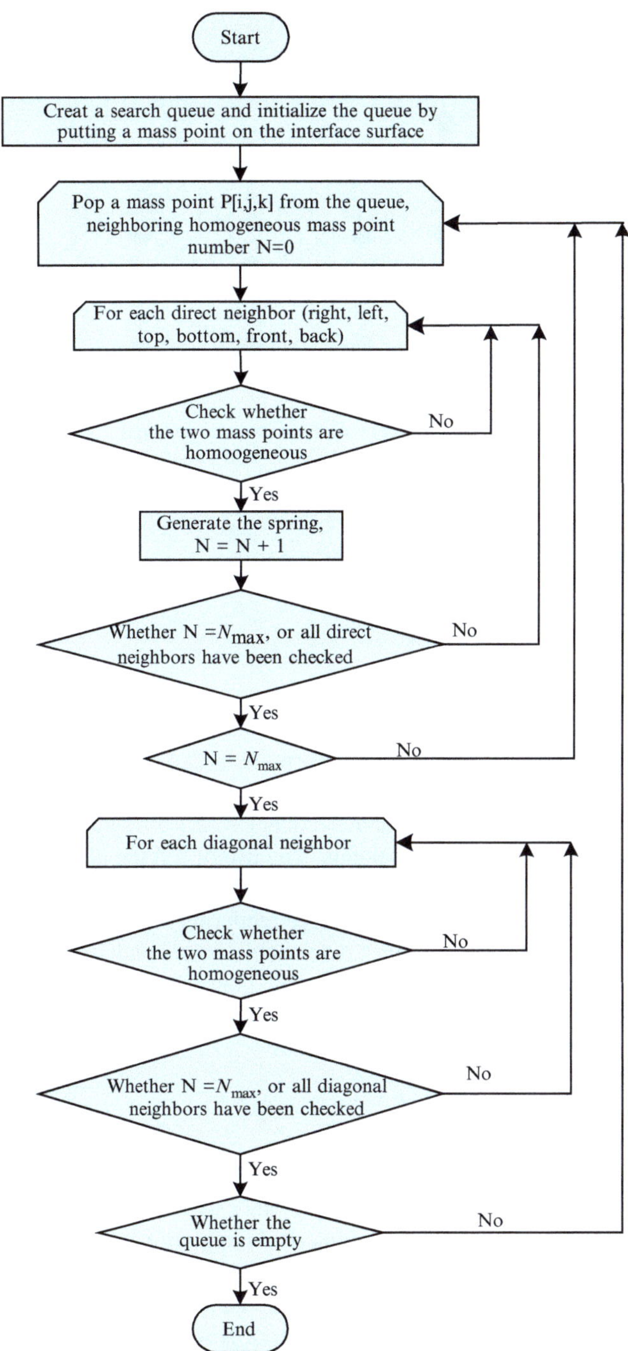

Fig. 2.9 Flowchart of interface marching algorithm

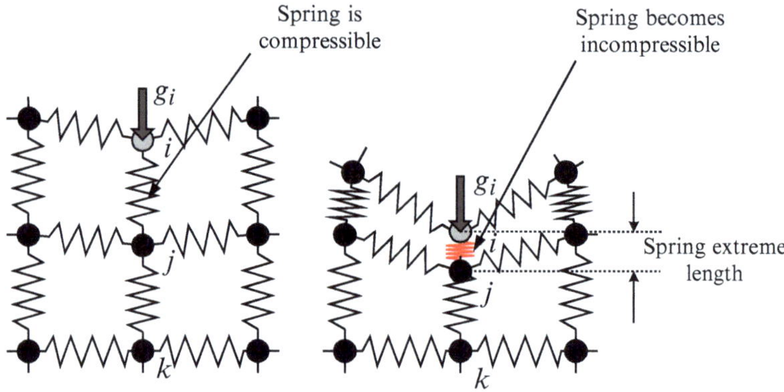

Fig. 2.10 Spring becomes rigid (incompressible) at its extreme length

$$\begin{cases} f_i = \sum\limits_{\{t \in n_i, t \neq j\}} \dfrac{k_{it} \Delta l_{it} \vec{d}_{it}}{|\vec{d}_{it}|} + f_j \\ f_j = \sum\limits_{\{t \in n_j, t \neq i, t \neq k\}} \dfrac{k_{jt} \Delta l_{jt} \vec{d}_{jt}}{|\vec{d}_{jt}|} + f_k \end{cases} \text{if } \Delta l_{ij} \geq \alpha_{\max} l_{ij} \text{ or } \Delta l_{ij} \leq \alpha_{\min} l_{ij}, \Delta l_{jk} \geq \alpha_{\max} l_{jk} \text{ or } \Delta l_{jk} \leq \alpha_{\min} l_{jk} \quad (2.5)$$

where n_i and n_j are the neighboring mass points of mass point i and j. f_k can be calculated recursively using the second formula in (2.5) if there are more extreme springs connecting to mass point k. A mass point touched by the haptic tool is defined as a contact mass point. Haptic feedback forces are calculated by interpolating spring forces of contact mass points. The use of a force accumulation method as well as appropriate material properties from heterogeneous deformable models can be used to provide realistic haptic force rendering for VR applications.

To find the equilibrium state of a deformable object under deformation forces, the governing force system of the deformable object is formulated first. The mass spring system of a deformable object is a dynamic system, which is governed by following second-order differential equations:

$$m_i \ddot{x}_i + d_i \dot{x}_i = g_i + f_i \quad (2.6)$$

where m_i, d_i are mass and damping coefficient of mass point i; $\ddot{x}_i$ and $\dot{x}_i$ are its acceleration and velocity, respectively; g_i is total external force and f_i is total internal spring force of mass point i. Several numerical integration schemes have been used to solve these Ordinary Differential Equations, such as explicit Euler method, fourth-order Runge–Kutta method and Verlet method [25,26]. All these numerical integration schemes suffer the same problem of slow convergence to system equilibrium, since each mass point may oscillate around its equilibrium position for a few times. A large integration time step h can accelerate convergence but may bring system instability.

In surgical techniques, it is often observed that dynamic behaviors of biological soft tissues cannot be readily determined by means of conventional surgical manipulation [26]. Biological soft tissues can reach their static equilibrium quickly. A quasi-static algorithm was used to find the static solution without considering inertia, mass, or damping [27]. However, this algorithm still needs to find a proper integration step experimentally to balance fast convergence and real-time performance.

In our earlier work presented in [19], a constrained local static integration method was developed for quick convergence and stable simulation. Static equilibrium of each moving mass point is calculated locally within its neighboring bounding box (NBB), as shown in Fig. 2.11. The NBB of mass point $[i, j, k]$ is defined by its six direct neighboring mass points (Fig. 2.11). Movement of each mass point is constrained within the NBB by using the following conditions:

$$\begin{cases} x_{[i-1,j,k]} \le x_{[i,j,k]} \le x_{[i+1,j,k]} \\ y_{[i,j-1,k]} \le y_{[i,j,k]} \le y_{[i,j+1,k]} \\ z_{[i,j,k-1]} \le z_{[i,j,k]} \le z_{[i,j,k+1]} \end{cases} \tag{2.7}$$

The local static equilibrium force method is applied only to aforementioned active mass point i shown as follows:

$$\sum_{j=0}^{n_i} K_{ij}[P_j - P_i - (P_{0,j} - P_{0,i})] = g_i \tag{2.8}$$

where P_i is the current position of mass point i to be calculated; K_{ij} is the spring coefficient of spring ij; P_j is the known current position of mass point j; $P_{0,i}$ and $P_{0,j}$ are the initial positions of mass points i and j; g_i is the external force on mass point

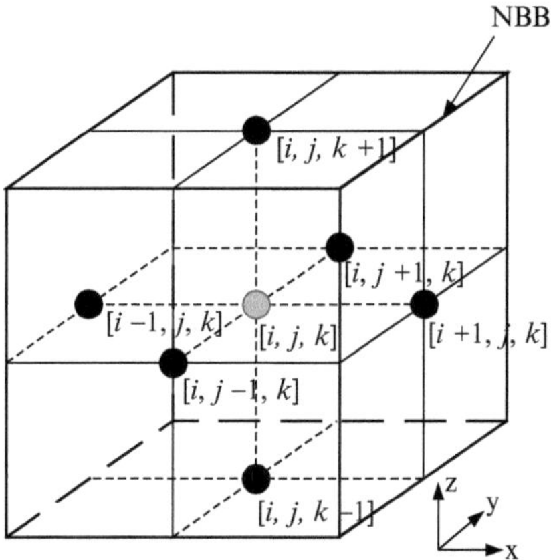

Fig. 2.11 Neighboring bounding box (NBB) of a mass point

i; and n_i is the number of the neighboring mass points of mass point i. The unknown position P_i in (2.8) can be expressed as follows:

$$P_i = \frac{-g_i + \sum_{j=0}^{n_i} K_{ij}[P_j - (P_{0,j} - P_{0,i})]}{\sum_{j=0}^{n_i} K_{ij}} \tag{2.9}$$

The static equilibrium for P_i is a local and temporary solution, which may be changed according to the dynamic perturbation of the nodes in the deformable models. For example, the variable P_j in (2.9) may be changed due to the dynamic movement of P_i resulting from the change in the local static equilibrium. In this paper, an iterative procedure is applied to (2.9) with the iteration index t shown as follows:

$$P_i^{t+1} = \frac{\sum_{j=0}^{n_i} K_{ij}[P_j^t - (P_{0,j} - P_{0,i})]}{\sum_{j=0}^{n_i} K_{ij}} \quad \text{where} = 0,1,2,... \tag{2.10}$$

During determination of the final equilibrium, each mass point moves toward its final solution, normally within a few iterations. When some active mass points only move very short distance from their last iterative locations and the responsive spring force is smaller than a predefined threshold value, these mass points are considered to have reached the final equilibrium status. When no more active mass points are present, the iteration is completed and the final equilibrium of the system is accomplished.

2.5 System Implementation and Examples

The presented techniques have been implemented at North Carolina State University on a 2.4 GHz dual-CPU workstation with 2 GB memory and a high-end graphics card, using Visual C++ .Net 2003 and OpenGL® library. The system is integrated with a 6-DOF (degree of freedom) Phantom® desktop haptic device, which provides 3-DOF force feedback. Figure 2.12 shows the lab setup of the implemented system with the 6-DOF haptic interface. An example heterogeneous deformable model was deformed by using the haptic device, as shown in Fig. 2.12. Multiple threads technique has been applied to deal with haptic rendering, graphics, and collision detection, respectively.

Figure 2.13 shows a procedure of using the presented techniques to construct the heterogeneous deformable model using medical imaging data. Using the presented techniques, a model of human right thigh is generated, as shown in Fig. 2.13. Skin, facial, femur, and nine types of muscles are integrated into one heterogeneous thigh

Fig. 2.12 Lab-setup of heterogeneous deformable modeling with a haptic force-feedback device

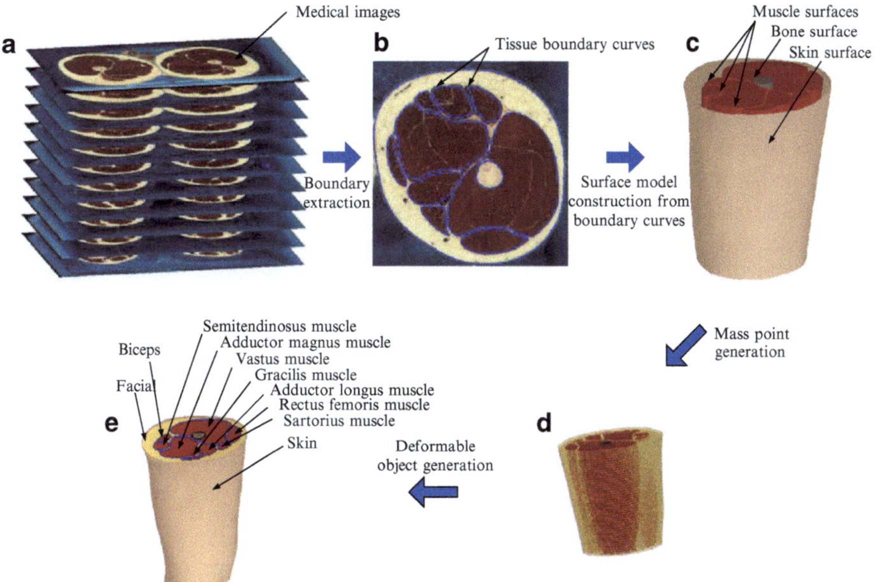

Fig. 2.13 Construction of a heterogeneous volumetric model from medical images

model. First, color medical images of different height sections are placed in a layer-by-layer manner as shown in Fig. 2.13a. The boundary curves of soft tissues and femur in the right thigh are sketched using splines as shown in Fig. 2.13b. From boundary curves on different layers, surface models of each tissue are constructed as shown in Fig. 2.13c. The tri-ray node snapping algorithm is then used to generate the mass points (Fig. 2.13d). During the mass point generation process, a flag is set for each mass point to indicate which tissue it belongs to. Figure 2.13e shows the final rendered heterogeneous deformable model.

Figure 2.14 shows a detailed profile of the deformation using a small compressive force via a haptic device. Several tissue layers including skin, subcutaneous fat, fascia, and muscles are involved in the deformation. The deformation is propagated from the model surface into the heterogeneous model. The upper layers exhibit larger deformation than deeper layers, as shown in Fig. 2.14. On the basis of our proposed techniques, no penetration happens between tissue layers during the deformation. The result is consistent with deformation behavior of actual soft tissues.

Figure 2.15a shows another heterogeneous deformable model of human anterior wall. This model includes 4,248 triangles on the skin layer and a total of 53,199 mass points inside the model. Tissue types are indicated by different colors, and interface surfaces between soft tissues are represented by bold lines. As shown in Fig. 2.15, a haptic tool is manipulated to push against the abdomen, similar to the interactive palpation on a patient's abdomen by medical examiners. As shown in Fig. 2.15a–c, an experiment is performed by pushing the model using the haptic tool along two paths with different depths. Figure 2.15d, e shows the resulting forces along the two paths. When the tool is moved just underneath the skin, there is no apparent force change because only fat tissues are deformed slightly (Fig. 2.15d). When the tool is pushed deep enough as shown in Fig. 2.15e, muscles are indirectly involved in the deformation process. Since muscles exhibit much higher stiffness values than skin or fat, a larger force is generated. As a result, the user can feel the existence of muscles by force feedback. Due to hand shaking during movement, the tool paths as well as the resulting forces demonstrate some fluctuation.

Figure 2.16 shows an example of haptic manipulation of a human right leg model, in which bone is involved. Bone is a rigid inorganic–organic composite material, which is generally considered to be incompressible. When soft tissues are gradually pushed against the bone surface, soft tissues will be compressed to their mechanical limit. Once the soft tissues can no longer be compressed, forces will increase dramatically if the user continues to push against the model. In this case, users can feel the bone using a haptic device.

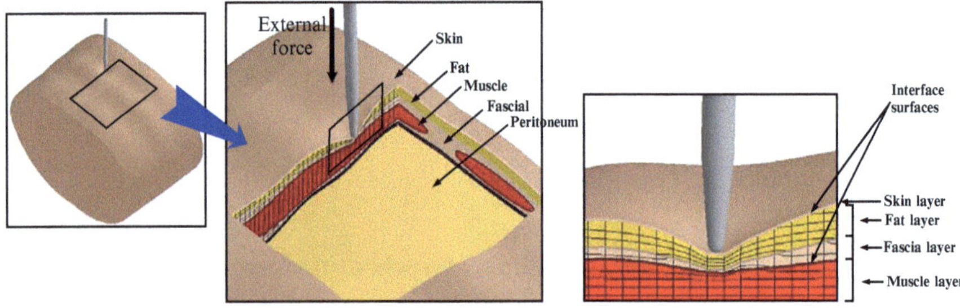

Fig. 2.14 Example of deformation on a heterogeneous anterior abdominal model via the haptic interface

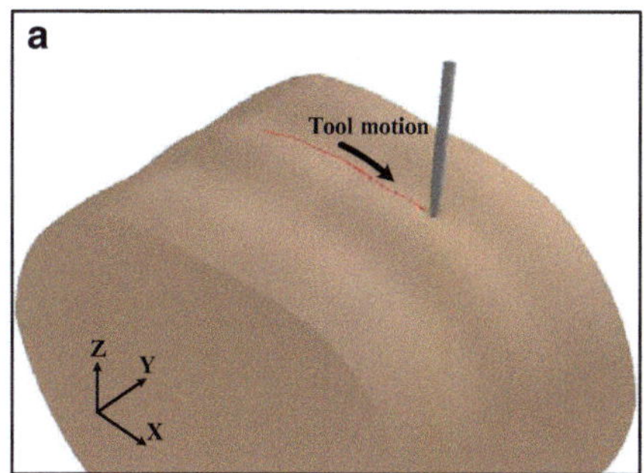

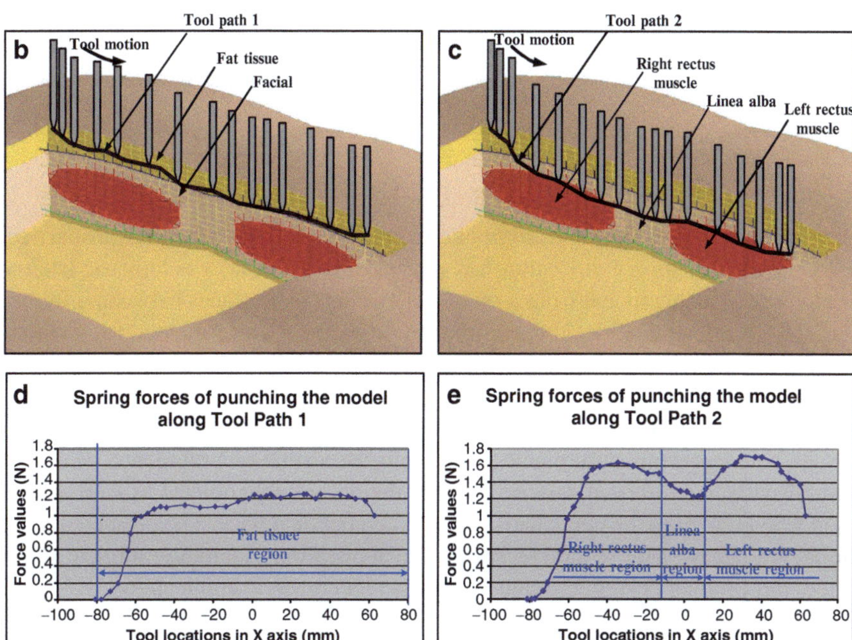

a Deformation of a model. **b** Fat tissues pressed along Path 1. **c** Muscles pressed along Path 2. **d** Response forces along Path 1. **e** Response forces along Path 2.

Fig. 2.15 Response forces via the haptic interface when pushing a tool along the different trajectories on the model. (**a**) Deformation of a model. (**b**) Fat tissues pressed along Path 1. (**c**) Muscles pressed along Path 2. (**d**) Response forces along Path 1. (**e**) Response forces along Path 2

2.6 Conclusions

In this paper, modeling of soft biological tissues and development of a haptic-based force-feedback interface for bio-manufacturing and medical applications have been presented. Detailed techniques for constructing heterogeneous deformable models were discussed for representing the internal structures of deformable biological

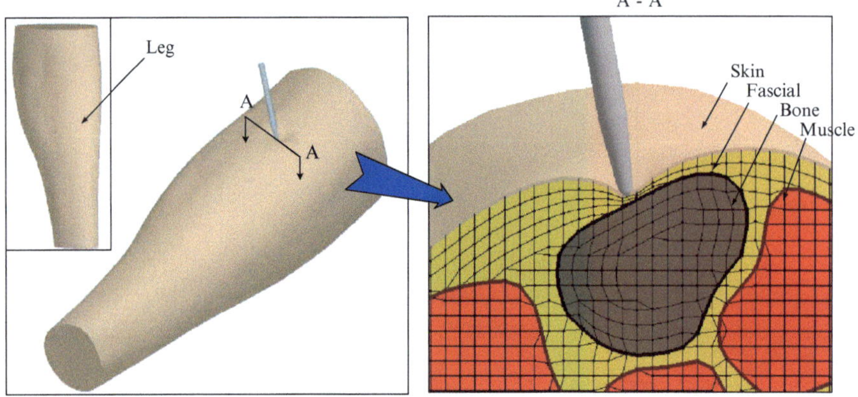

Fig. 2.16 Deformation on an example of human leg model with rigid bones via the haptic force feedback interface

objects that possess multiple components and nonuniform material properties. A tri-ray node snapping algorithm was presented to generate a volumetric heterogeneous deformable model from a set of object interface surfaces between different materials. A constrained local static integration method was presented for simulating deformation and accurate force feedback for heterogeneous structures. The presented heterogeneous deformable object modeling and haptic rendering can greatly enhance the realism and fidelity of virtual reality environments. By integrating the heterogeneous deformable model into the virtual environments, users can both visually see different materials inside the deformable objects as well as feel them when touching the deformable object via a haptic device. The presented techniques may be used for several technological applications, including surgical simulation, bio-product design, bio-manufacturing, and medical applications.

Acknowledgment This work was partially supported by the National Science Foundation (NSF) Grants (DMI-0300297, CMMI-0553310) to North Carolina State University. Their support is greatly appreciated.

References

1. Akagi Y, Kitajima K (2006) Computer animation of swaying trees based on physical simulation. Comput Graph 30(4):529–539
2. Al-khalifah A, Roberts D (2004) Survey of modeling approaches for medical simulators. In: Proceedings of 5th Intl Conf. Virtual Reality & Assoc. Tech., Oxford, UK, pp 321–329
3. Dewaele G, Cani M-P (2004) Interactive global and local deformations for virtual clay. Graph Models 66:352–369
4. Lin S, Lee Y-S, Narayan R (2008) Heterogeneous material modeling and virtual prototyping with 5-DOF haptic force feedback for product development. Int J Mechatronics and Manuf Syst 1(1):43–67

5. Lin S, Lee Y-S, Narayan R, Shin H. Bio-tissues modeling and interface development for bio-manufacturing and medical applications. In: Proceedings of 2007 Industrial Engineering Research (IERC) conference, Nashville, TN, 19–23 May 2007, pp 281–286
6. Georgii J, Westermann R (2005) Interactive simulation and rendering of heterogeneous deformable bodies. In: VMV 2005, Erlangen, Germany, 16–18 Nov 2005
7. Sarni S, Maciel A, Boulic R, Thalmann D (2004) Evaluation and visualization of stress and strain on soft biological tissues in contact. In: Proceedings of shape modeling conference, pp 255–262
8. Desbrun M, Schroder P, Barr A (1999) Interactive animation of structured deformable objects. In: Proceedings of 1999 conference on graphics interface, pp 1–8
9. Mendoza C, Laugier C (2003) Tissue cutting using finite elements and force feedback. In: Proceedings of the international symposium on surgery simulation and soft tissue modeling, pp 175–182
10. Bourguignon D, Cani M (2000) Controlling anisotropy in mass-spring systems. In: Proceedings of the 11th eurographics workshop, Interlaken, Switzerland, Aug 2000, pp 21–22
11. Picinbono G, Delingette H, Ayache N (2003) Non-linear anisotropic elasticity for real-time surgery simulation. Graph Models 65:305–321
12. Ren Y, Lai-Yuen SK, Lee Y-S (2006) Virtual prototyping and manufacturing planning by using tri-dexel models and haptic force feedback. Virtual Phys Prototyping 1(1):3–18
13. Zhu W, Lee Y-S (2005) A visible sphere marching algorithm of constructing polyhedral models from dexel models for haptic virtual sculpting. Robot Comput Integrated Manuf 21(1):19–36
14. Lin S, Lee Y-S, Narayan R (2007) Collaborative haptic interfaces and distributed control for product development and virtual prototyping, In: Proceedings of ASME manufacturing science and engineering conference. ASME MSEC2007-31214
15. Peng X, Zhang W, Leu MC (2006) Freeform modeling using sweep differential equation with haptic interface. Virtual Phys Prototyping 1(3):183–196
16. Ye J, Campbell RI (2006) Supporting conceptual design with multiple VR based interfaces. Virtual Phys Prototyping 1(3):171–181
17. Gibson I (2006) Rapid prototyping: from product development to medicine and beyond. Virtual Phys Prototyping 1(1):31–42
18. Basdogan C, De S, Kim J, Muniyandi M, Kim H, Srinivasan MA (2004) Haptics in minimally invasive surgical simulation and training. IEEE Comput Graph Appl 24(2):56–64
19. Lin S, Lee Y-S, Narayan R (2007) Heterogeneous soft material modeling and virtual prototyping with 5-DOF haptic force feedback for product development, In: Proceedings of 2007 international conference on advanced research in virtual and physical prototyping (VRAP2007), VRAP: 187–193
20. Fung YC (1993) Biomechanics: mechanical properties of living tissues, 2nd edn. Springer-Verlag, New York
21. Meier U, Lopez O, Monserrat C, Juan MC, Alcaniz M (2005) Real-time deformable models for surgery simulation: a survey. Comput Methods Programs Biomed 77:183–197
22. Mollemans W, Schutyser F, Cleynenbreugel JV, Suetens P (2003) Tetrahedral mass spring model for fast soft tissue deformation. In: Proceedings of surgery simulation and soft tissue modeling: international symposium, IS4TM, pp 145–154
23. Samanta K, Koc B (2005) Feature-based design and material blending for free-form heterogeneous object modeling. Comput Aided Des 37:287–305
24. Nakao M, Kuroda T, Oyama H, Sakaguchi G, Komeda M (2006) Physics-based simulation of surgical fields for preoperative strategic planning. J Med Syst 30(5):371–380
25. Bielser D, Maiwald VA, Gross MH (1999) Interactive cuts through 3-dimensional soft tissue. Comput Graph Forum 18(3):31–38
26. Sørensen TS, Mosegaard J (2006) An introduction to GPU accelerated surgical simulation. In: The 3rd symposium on biomedical simulation, Zurich, Switzerland, lecture notes in computer science (series No. 4072), pp 93–104
27. Brown J, Sorkin S, Bruyns C, Latombe JC, Montgomery K, Stephanides M (2001) Realtime simulation of deformable objects: tools and application. In: Computer animation, Seoul, Korea, 7–8 Nov 2001, pp 228–236

Chapter 3
Computer-Aided Process Planning for the Layered Fabrication of Porous Scaffold Matrices

Binil Starly

3.1 Introduction

Rapid Prototyping (RP) technology promises to have a tremendous impact on the design and fabrication of porous tissue replacement structures for applications in tissue engineering and regenerative medicine. The layer-by-layer fabrication technology enables the design of patient-specific medical implants and complex structures for diseased tissue replacement strategies. Combined with advancements in imaging modalities and bio-modeling software, physicians can engage themselves in advanced solutions for craniofacial and mandibular reconstruction. For example, prior to the advancement of RP technologies, solid titanium parts used as implants for mandibular reconstruction were fashioned out of molding or CNC-based machining processes (Fig. 3.1). Titanium implants built using this process are often heavy, leading to increased patient discomfort. In addition, the Young's modulus of titanium is almost five times that of healthy cortical bone resulting in stress shielding effects [1,2]. With the advent of CAD/CAM-based tools, the virtual reconstruction of the implants has resulted in significant design improvements. The new generation of implants can be porous, enabling the in-growth of healthy bone tissue for additional implant fixation and stabilization. Newer implants would conform to the external shape of the defect site that is intended to be filled in. More importantly, the effective elastic modulus of the implant can be designed to match that of surrounding tissue. Ideally, the weight of the implant can be designed to equal the weight of the tissue that is being replaced resulting in increased patient comfort. Currently, such porous structures for reconstruction can only be fabricated using RP-based metal fabrication technologies such as Electron Beam Melting (EBM), Selective Laser Sintering (SLS®), and 3D™ Printing processes.

Porous scaffold structures also play a critical role in the field of tissue engineering and regenerative medicine. Scaffold structures simulate extracellular matrices onto which living cells can attach, grow, and form new tissues for eventual diseased

B. Starly (✉)
School of Industrial Engineering, University of Oklahoma, Norman, OK, 73019, USA
e-mail: starlyb@ou.edu

R. Narayan et al. (eds.), *Printed Biomaterials*, Biological and Medical Physics, Biomedical Engineering, DOI 10.1007/978-1-4419-1395-1_3,

Fig. 3.1 Titanium mandible implant fabricated using a CNC machining process

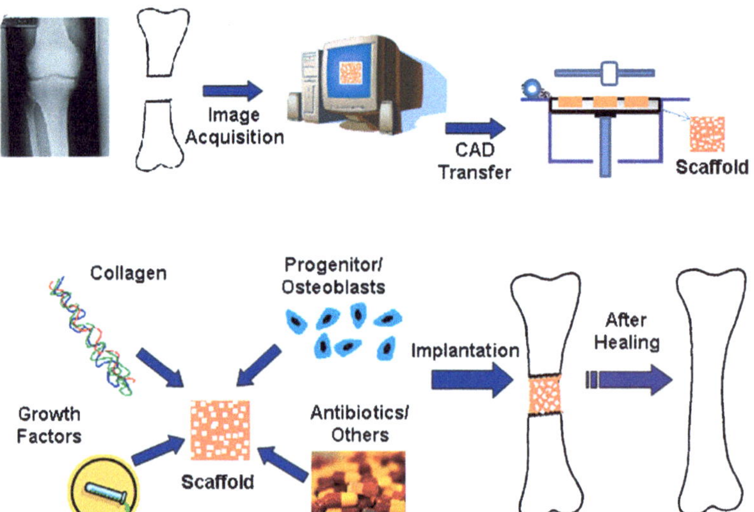

Fig. 3.2 Scaffold guided tissue engineering with porous structures built using RP-based systems

tissue replacement (Fig. 3.2) [17–20]. In the growth and migration processes that give rise to tissue function and morphogenesis, transplanted cells are influenced by a wide variety of factors including their own genetic mechanisms, the cell's bio-molecular composition, and the scaffold architecture. The architecture directs and induces cell proliferation, migration, and differentiation [3]. When designing load bearing scaffolds for bone and cartilage tissue application, the scaffolds are designed with intricate porous architecture to exhibit appropriate mechanical properties, biocompatible surface chemistry, facilitate transport of nutrients, and provide a conducive environment for cell and tissue in-growth [4]. However, designing the micro-architecture for these scaffold structures is limited due to inadequate CAD technologies with regards to the representation of such complex micro-structures and the inability to transfer microstructure geometric information to RP machines. The micro-structures contain spatially oriented pores that guide the overall placement and growth of cells. Uniformly placed pores enable the even distribution of cells for appropriate tissue growth and sustenance.

The interior architecture of these scaffolds is designed as a pattern of extrusions, cuts, and holes across the surface of the scaffolds in a CAD platform. Alternatively, the slice layers for the 3D structure are filled in with contours or roads to generate the interior architecture. The former method is used by 3DP™-, SLS®-, and SLA®-based RP machines for scaffold fabrication while the latter method is used primarily by FDM and Micro-nozzle extrusion-based machines. [5] used an assembly of designed unit cells and then intersected with the scaffold external structure to obtain the interior architecture using a Pro-E-based CAD environment. [6] have used a similar approach in defining the interior architecture where the whole process was performed using STL-based unit cell models. After final design, the model is converted to the STL format as a means of data transfer to RP machines for fabrication. A sample boolean operation to generate porous implants is shown in Fig. 3.3 [15,16,21]. However, this method has the following limitations:

1. STL tessellation involves approximation of surfaces with triangular facets that are incapable of representing the scaffold's finer micro-level pore feature dimensions and intricate internal architectures. For example, as model precision demands become more stringent, the number of facets required to adequately approximate the model will increase. This usually ends up having computing inefficiencies with the STL file due to the extremely large file sizes. These STL files may also contain triangular facet errors which then will require file repair tools to handle them.

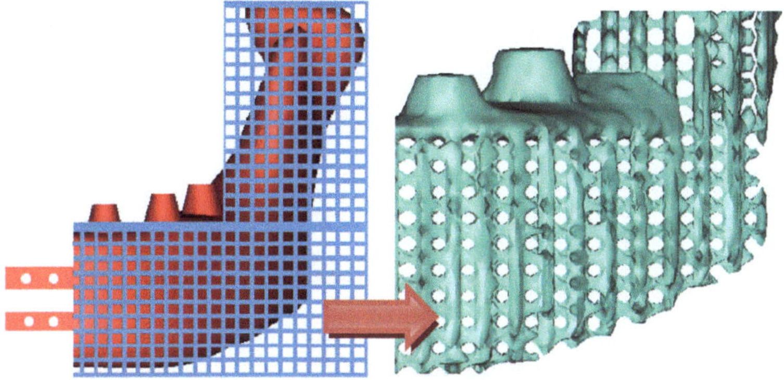

Fig. 3.3 Boolean operation between the replacement structure and a porous template to generate the desired porous implant

2. Database in the STL format cannot include other design intents within the model, for example, topology or internal material variations, and nonuniform spatial distribution etc. Therefore, heterogeneous representation is not allowed within the STL format.
3. Expensive computing power is required when the number of features and cuts in the scaffold become large enough to cause a memory overhead for the CAD software used. This would mean that while RP systems are generally capable to create pores in the 100–300 μm, commercially available CAD systems are not generally capable to handle such complex structures.

This chapter will describe an innovative pattern mapping technique to develop process tool paths for the freeform fabrication of porous tissue replacement structures from characteristic architecture patterns of the scaffold building blocks. The output of such a process would be to have scaffolds with intended interior architecture to ensure the right pore size, pore shape, and interconnectivity throughout the scaffold while making use of currently available CAD technologies. The interior architectures of these scaffolds are not limited to a single design pattern but extended for multiple interior patterns within the same structure. This method would be absolutely critical for the design of the next generation of patient-specific implants and tissue scaffolds enabled by RP technologies.

3.2 Internal Architecture Design of Tissue Scaffolds

Although we can use 3D models to represent individually designed unit cells, CAD is not generally capable of representing a unit cell assembly, for example, using different unit cells to fill in the femur structure as shown in Fig. 3.4. Since each unit cell is designed with a complex set of features, the combination of hundreds of unit cells will result in the formation of an intricate scaffold internal architecture which is beyond the capability of many currently available commercial CAD systems. In addition, the STL format which serves as the current industry standard for the data exchange between CAD and RP systems cannot be used to efficiently transfer data for such complex intricate structures.

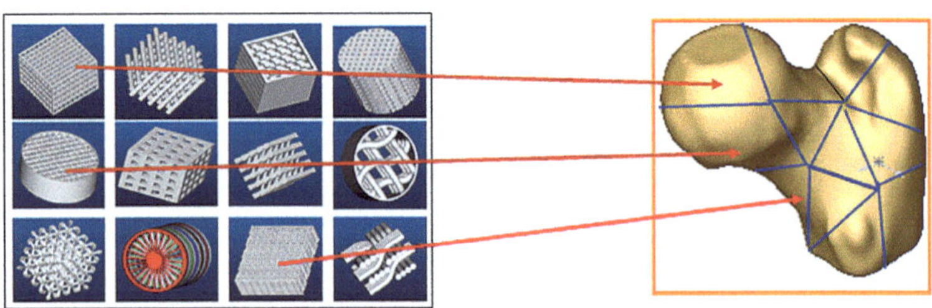

Fig. 3.4 Unit cell assignment within Scaffold structure

Therefore, new methods to transfer the information of scaffolds with a highly complex interconnected porous structure to a process tool path data for final fabrication must be developed. We have devised a direct pattern mapping solution to generate process-planning instructions for the freeform fabrication of tissue scaffolds and porous implants. The novelty of this approach is that it allows the generation of the scaffold process-planning instruction set without the explicit creation of a unit cell filled 3D CAD scaffold model, thus avoiding the difficulty of using CAD to represent the assembly of the unit cell models. The approach stems from the principle of the decomposition and material accumulation of layered manufacturing principles [7]. The approach uses the characteristic patterns to represent the individual unit cell 2D interior architectures, and assembles the individual characteristic pattern to form a 2D layered pattern for the scaffold. The processing tool path is generated for every layer based on the layered scaffold pattern, and is then applied to instruct RP machines to build 3D scaffold layer by layer. In this way, it is unnecessary to create an explicit 3D CAD-based scaffold model, but with the implicit representation be able to design and control the internal architecture of the scaffold. Detailed development of the approach is outlined in the section as follows.

3.3 Process Methodology

3.3.1 Step 1: Determine the Layered Processing Plane ((x,y,z))

The 3D volumetric scaffold V is sliced into layers during the model decomposition, and stacked back with materials during the material accumulation in an ordered sequence. The sequence is a number dispersal in three-dimensional space V in which the layered manufacturing process is realized. Let us introduce a finite discontinuous real number set C to indicate the discrete layers through which V is decomposed:

$$C = \{c_k; a \le c_1 < c_2 < \ldots < c_{k-1} < c_k < c_{k+1} < \ldots < c_n < c_{n+1} \le b, k = 1, 2, \ldots, n\} \quad (3.1)$$

where subscript k represents the layer index in a total n layers, and constants a and b represent the values of lowermost and uppermost sequence for a particular object V. Therefore, corresponding to each C_k, we can define a sequence function $\varphi_k(x,y,z)$, and assume it is equal to the ordered number c_k, as shown in Fig. 3.5, we thus define a series of planes in V:

$$\varphi_k(x, y, z) = c_k, \quad k = 1, 2, \ldots, n \quad (3.2)$$

We call these planes, the layered processing planes and represent them by a set of scale function $\varphi_k(x,y,z)$, with $k=1,2,\ldots,n$. It is important to note that the layered processing plane $\varphi_k(x,y,z)$ represents both the indication of the processing layer, and the layered exterior geometry represented by the CAD model for the scaffold. In the model decomposition process, the layered processing planes intersect and slice the designed 3D scaffold. In the material accumulation process, the layered processing planes form the processing layers on which material will be added. The concept of the decomposition and accumulation, and the layered processing plane is further depicted in Fig. 3.5b.

On each of the defined layered processing plane $\varphi_k(x,y,z)$ in the volumetric scaffold, we define S^k as a 2D layered scaffold pattern, which can be considered as a union of the partitioned unit cell characteristic patterns:

$$S^k = S_1^k \cup S_2^k \cup S_3^k \cup \ldots \cup S_i^k \tag{3.3}$$

For simplicity, we assume that the thickness of the slicing layer for unit cells and for the scaffold are uniform and equal in magnitude. Slicing algorithms based on indirect or direct techniques [8,9] can be employed to obtain the external contours $(\varphi_k)(x,y,z))$ of the porous structure and the selected unit cell architecture.

3.3.2 Step 2: Generate Toolpath Based on the Direct Pattern Mapping Technique

The layered processing toolpath for freeform fabrication is then generated based on the 2D layered scaffold pattern. The above process can be briefly summarized into the following major steps:

1. Define subvolume V_i (unit cell) and the spatial position P_i; the discrete layers of V, layered thickness, and the layered processing plane $\varphi_k(x,y,z)$ for scaffold exterior geometries (as defined in Step 1).

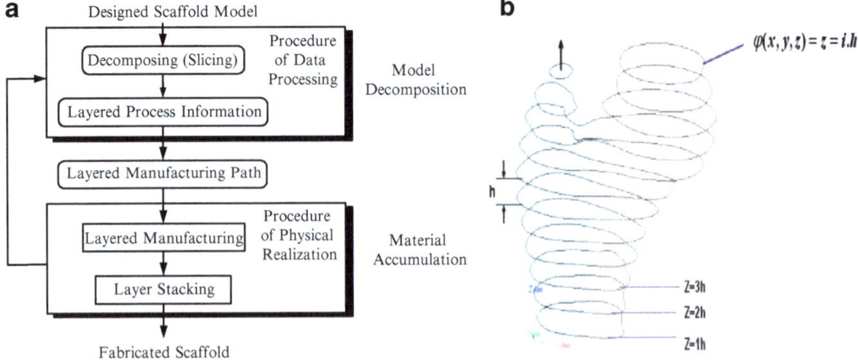

Fig. 3.5 Decomposition and accumulation for layered manufacturing

2. Determine the unit cell characteristic patterns S_i^j ($i=1,2,\ldots,m$, and $j=1,2,\ldots$) by slicing them.
3. Form a 2D layered scaffold pattern S^k ($k=1,2,\ldots,n$) (or layered Interior Pattern) for a given scaffold sliced layer by a union of all the unit cell patterns.
4. Perform a intersection Boolean operation $S^k \varphi_k(x, y, z)$ between the 2D layered scaffold pattern and the exterior scaffold geometry given by ϕ to remove the unwanted regions of the scaffold sliced layer.
5. Conversion of the Interior Scaffold Pattern to process tool path information for freeform fabrication.
6. Repeating Step 3 to Step 5 till $k=n$.

We use a square block type unit cells with open pore in the center as shown in Fig. 3.6 as an example to illustrate the above process. Assume that the designed scaffold consists of "m" unit cells with four unit cells being assigned to the top of the scaffold volume (Fig. 3.6). Assuming that the number of slices of the scaffold has been determined, the layered processing planes $\varphi_k(x,y,z)$ ($k=1,2,\ldots,n$) for the scaffold exterior geometry are known, and the characteristic patterns of the unit cells S_i^j corresponding to the $\varphi_k(x,y,z)$ have been computed and available in the database. The unit cell id P_i within different regions on this layer is used as a key to retrieve the appropriate unit cell characteristic patterns. A scenario has been illustrated as shown in Fig. 3.7 by taking the case of two scaffold slice layers shown as slice layer 1 and slice layer 2. The 2D layered scaffold patterns for scaffold layer 1, S^1, and layer 2, S^2, are generated by the union of four unit cell characteristic patterns S_i^1 and S_i^2 ($i=1,2,\ldots,4$), respectively and are shown in the right side of Fig. 3.7. Printing maps consisting of the geometric raster pattern for the slice layer 1 and slice layer 2 are then generated based on S^1 and S^2.

By applying intersection Boolean operation between the Interior Scaffold Pattern and the layered processing plane $\varphi_k(x,y,z)$, we then obtain the layered manufacturing maps which will be used for the process tool path generation. The tool path can be generated based on the particular RP process used. The calculated

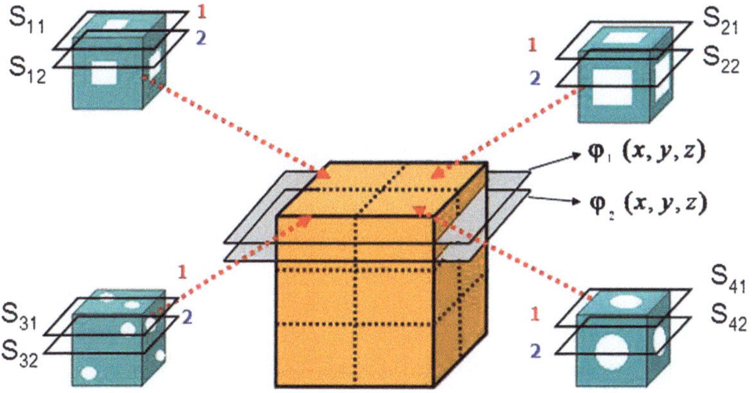

Fig. 3.6 Slicing of unit cell and Scaffold and associating them to specific regions

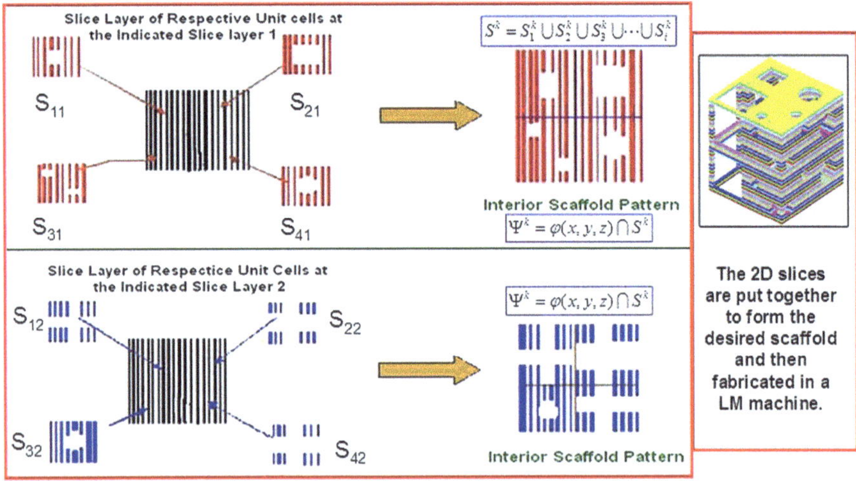

Fig. 3.7 Intersection of 2D slice patterns onto Scaffold 2D pattern to form 3D structure

intersection points will serve as the starting point for the tool path information for a 3DP™-based RP machine. For vector contour-based machines, the same scaffold pattern will be used to generate contour vectors by joining the intersection points in a linear loop technique and forming closed loop vector tool paths. The data flow chart is shown in Fig. 3.8.

3.4 Design Manufacturing Interface

A sample process-planning scenario to design and fabricate tissue scaffold structures is depicted in Fig. 3.9. In the custom implant planning stage, the CT/MRI images of the defected region are obtained and reconstructed by image reconstruction techniques [10]. Through reverse engineering techniques, a CAD model of the region of interest is made. Simultaneously, mechanical characterization techniques would be employed to obtain the mechanical properties of the region to be replaced. On the basis of the characterization process and biological requirements, an appropriate unit cell is designed and selected to fill into the desired scaffold structure. The CAD file is then converted to a STEP file format with intended geometrical and material information. Using an appropriate 3D modeling kernel, the scaffold implant reconstructed from the STEP definitions is then sliced depending on the slice parameters set using the slice module of the software. The slice module based on the direct or indirect slicing approach can be used. The slicing module defines the slice layers as scan raster lines to support the open-end architecture that is required for the scaffold structures. Once these scan lines are defined, complete information that consists of the slice layers as well as the material information is stored in a print job database that serves as a temporary storage as well as for future retrieval. The scan lines are converted to machine code instructions, which are then sent to the machine for fabrication.

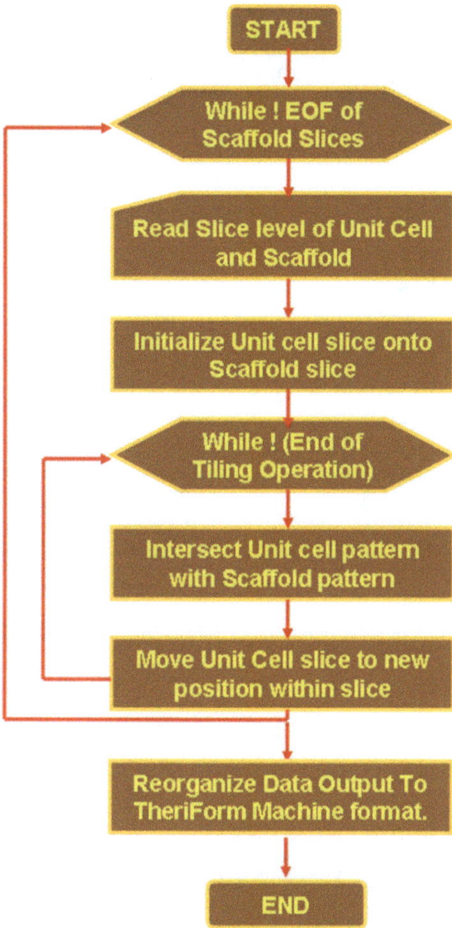

Fig. 3.8 IAD algorithm

The steps and algorithms outlined above are implemented within a fabrication planning software to control the different parameters of the fabrication process. The algorithms are implemented in Microsoft's .NET development environment using C# as the programming language. Figure 3.10 gives the different component modules with the functional flow of data within the fabrication preprocessing environment.

3.5 Case Study Models

The developed algorithm was tested for seven different kinds of models each with certain characteristics of its own to test the capability of the algorithm. The algorithm was tested with a couple of simple to complex models as shown in Figs. 3.11–3.17. The size of the selected unit cell is arbitrarily selected to be a 5 mm cube with

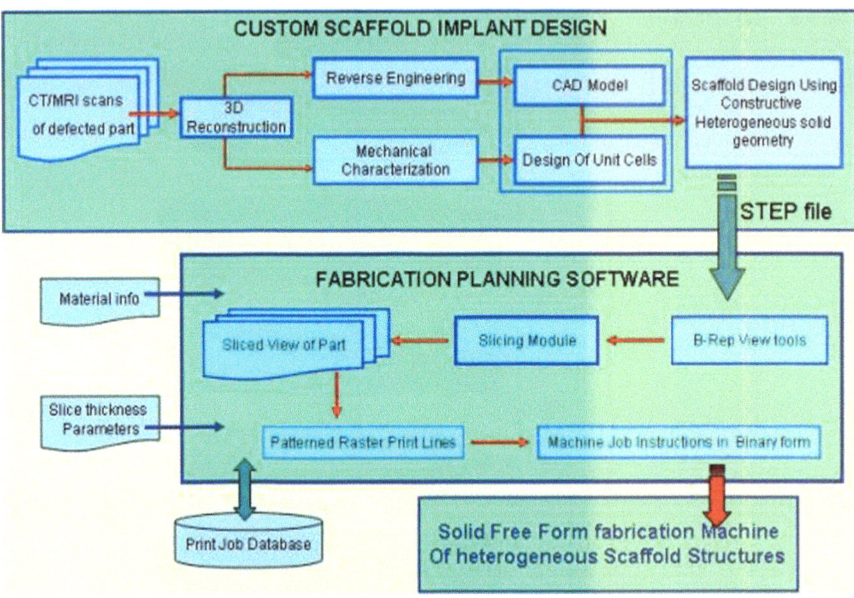

Fig. 3.9 Flowchart of process planning for fabricating tissue scaffold

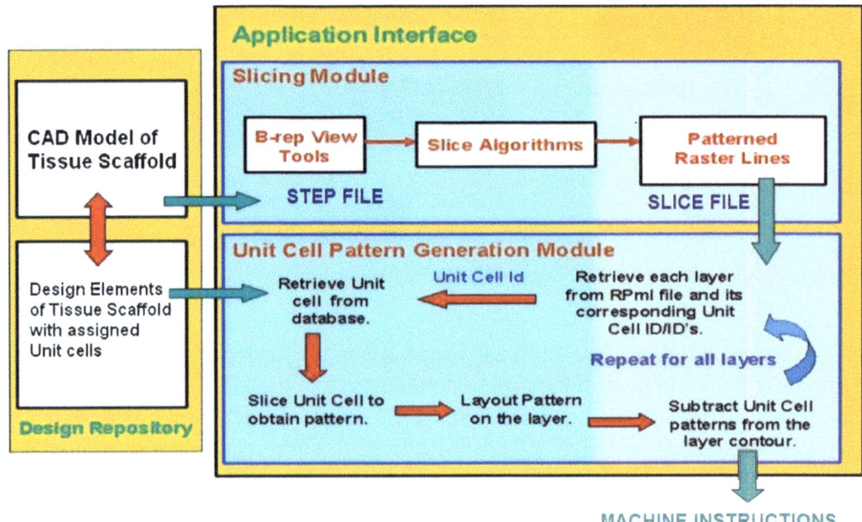

Fig. 3.10 Fabrication planning modules

characteristic shapes of either a square pore or a cylindrical hole within them. The blocks define the outer shape of the scaffold while the unit cells define the interior architecture of the scaffold. Both the blocks and the unit cells were sliced at equal intervals and stored in files. Given below are the test case results and its various process parameters.

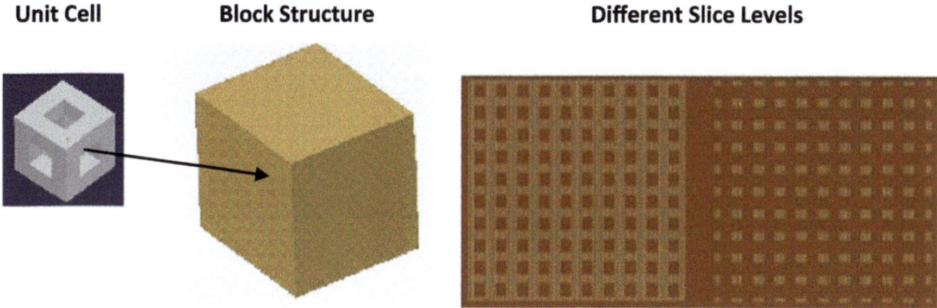

Fig. 3.11 (**a**) Square pore unit cell architecture with 5 mm×5 mm×5 mm unit cell with 2.5 mm×2.5 mm×2.5 mm pore size; (**b**) Square Block scaffold structure; (**c**) two different slice levels showing interior pattern structure

3.6 Model #2: Circular Unit Cell Assigned to a Cylinder scaffold

In this model, circular pore holes have been tested. As shown in Fig. 3.12, the circular patterns have been mapped into the cylindrical scaffold in a well uniform manner. The culling of the unit cell on the outer edges of the model is important to ensure proper interconnectivity within the scaffold.

3.7 Model #3: Circular Unit Cell Assigned to a Bone Architecture Scaffold

This model case study tests the ability of the algorithm to position itself correctly when the scaffold slice levels change. The results displayed in Fig. 3.13 show the pattern at four different slice positions within the final scaffold. The unit cell positions do not change and always ensure a through hole connectivity within the scaffold.

3.8 Model #4: Multiple Unit Cell Architecture Assigned to the Bone Structure

In this case study, the bone had been divided into four different sections. Each section is grouped together to form an assembly and is assigned a specific unit cell. On the basis of the id that each section has been assigned, during the mapping stage, the appropriate ID is retrieved and patterned onto the slice file. Figure 3.14 shows the four different interior architectures present at different slice levels of the bone structure.

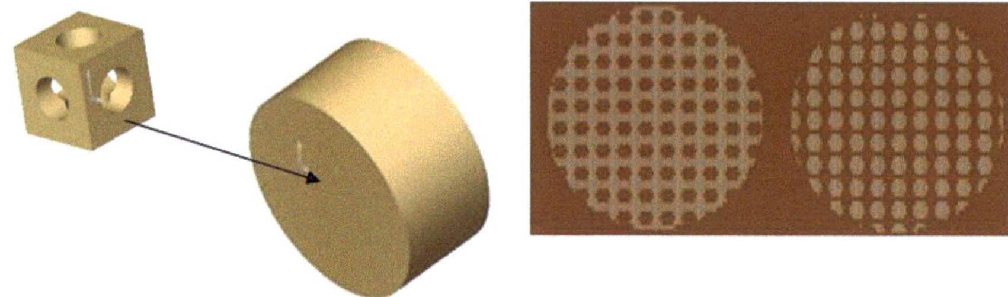

Fig. 3.12 (**a**) Unit cell with circular pore architecture – 5 mm×5 mm×5 mm unit cell with a 2.5 diameter 2.5 mm pore size; (**b**) cylindrical scaffold model structure; (**c**) circular holes patterned within interior at different slice levels

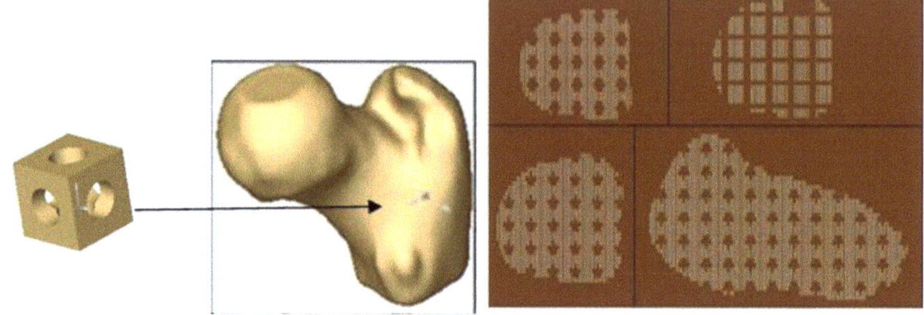

Fig. 3.13 (**a**) A circular pore architecture unit cell structure; (**b**) irregular model structure such as Bone; (**c**) four different slice levels of the bone scaffold structure showing circular holes

3.9 Model #5: Multiple Patterns Within a Single Slice

In this case study, we have two different unit cell architectures that are assigned to a single slice layer. As seen in Fig. 3.15, a circular unit cell has been assigned to the cylindrical subregion of the scaffold, while a square scaffold has been assigned to the rectangular block region of the scaffold. As seen in Fig. 3.15, multiple patterns exist within the single slice layer. Also note the sudden change in architecture at the interface region of the two subregions.

3.10 Model #6: Sweep Regions Within Scaffold

This model shown in Fig. 3.16 particularly shows how complex internal architectures can be achieved quite easily with no complicated CAD operations. Such internal architectures are hard to achieve in conventional CAD software. Two unit cell architectures have been assigned to sweep like subregions within the scaffold.

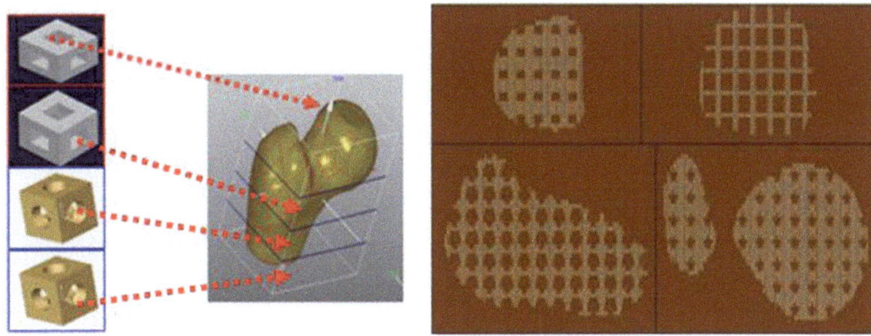

Fig. 3.14 (**a**) Bone structure divided into four; different levels with four unique unit cell structures assigned; (**b**) four different slice levels showing four different unit cell patterns

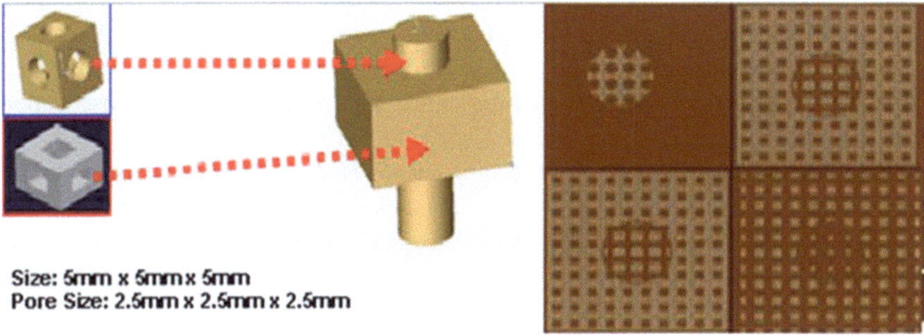

Fig 3.15 Two different pore architectures assigned to a model structure with multi patterns in a single layer

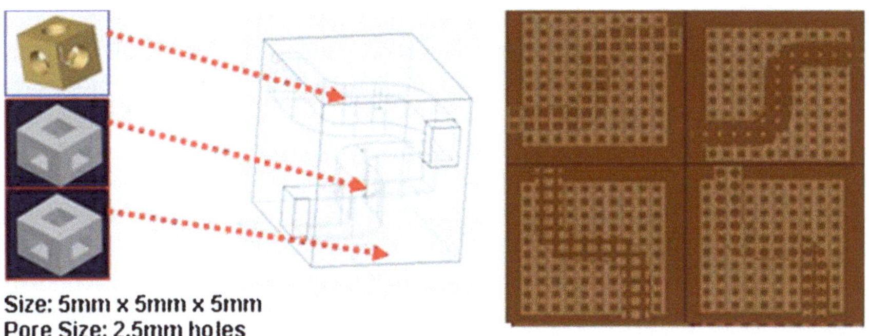

Fig. 3.16 Three pore architectures assigned to a shaped scaffold

3.11 Model #7: Area Selection Criterion

This particular model shows a unit cell architecture assigned to a cone model. An area selection criterion has been applied. If the area of the scaffold slice layer falls within a preset value, patterning does not take place and is left alone. This type of

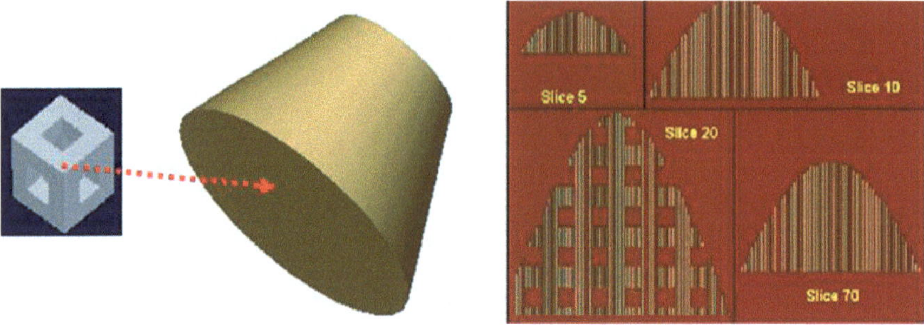

Fig. 3.17 Slices 5, 10, 70 do not meet the area selection criterion and hence no patterning is observed

scenario may be useful when considering the structural integrity of small features within the scaffold. The algorithm, however, does not work for multiple small features within the model. It calculates the entire area of the slice level based upon which decision to pattern or not takes place (Fig. 3.17).

The developed algorithm does not require any huge memory requirements and its run time only depends on the number of slices that make up the block. The size of the unit cell can go further smaller and its size is selected depending on other factors such as the desired pore size, the capability of the selected RP machine, and material particle size.

3.12 Fabrication Using 3DP™ Systems: TheriForm™ Technology

Models in Figs. 3.12 and 3.13 have been fabricated using the TheriForm™ fabrication machine. The principle of operation works similar to that of any 3DP™-based RP machine and is shown in Fig. 3.18. The powder bed consists of a biomedical grade material which is fused together by heat and a binder solution. Each layer in the build bed is incrementally pushed up by a fixed slice thickness depending on the type of material used, surface finish and accuracy needed, functionality of the required product. The nozzle through which the binder ejects is controlled by an x–y positioning system which is in turn controlled by the machine instructions generated from the models to be fabricated. In this case, the algorithm generates the machine level instructions for fabrication of the internal micro-architecture of the models. Each entry and exit point defined by the ray instructs the machine to start and stop dropping the binder respectively. Deposition of binder in one complete layer defines the material space for the model in its current layer and the process is repeated until all the layers have been fabricated.

Figure 3.18a, b shows a cylindrical shaped and a bone scaffold designed using the methodology and fabricated using the TheriForm™ machine. The scaffolds were made out of alumina with a slice thickness of 0.48 mm and sintering temperature at 80°C. As can be seen, the scaffolds have the required internal architecture as defined by the unit cell. A CAD model of the entire model with all the complete pores in them does not exist but merely an association is made.

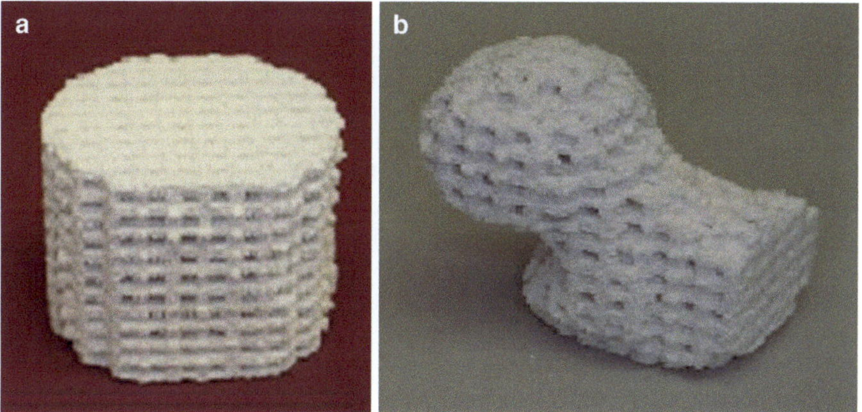

Fig. 3.18 Cylindrical and bone shaped scaffold fabricated using the Theriform™ RP system

Redefinition of the unit cell model architecture results in a different interior structure for the scaffold models.

3.13 Conclusions

The developed methodology facilitates the creation of more advanced scaffold designs that cannot be achieved using conventional CAD since each layer can individually be designed to have the desired layer pattern. This pattern is obtained from the assigned unit cell to the scaffold outer architecture to form the desired scaffold. With the combination of all layers put together, the selected interior unit cell architecture is fabricated within the scaffold structure. The advantages of this design process include:

- The developed approach can help in the design of scaffolds with complex interior architectures which by current techniques are impossible to create using CAD software or other 3D software. The developed approach does not exclude the use of commercial CAD software but uses it as an ancillary tool in the design of scaffolds with macro and micro architecture.
- The interior architecture of scaffolds can be designed and controlled along with the provision of multiple material specifications by associating the information to the unit cell. During the pattern layout stage, the information is translated to instruction set that can recognize the different materials regions.
- Direct transfer of CAD model data to RP machines avoids the bottleneck inherent in the use of the STL format.
- The design of outer scaffold shape and the design of its interior are separated and hence do not present a memory overhead for CAD software. By separating their designs, new architectures with varying properties can be designed and hence aid in the fabrication of better scaffolds for tissue engineering applications.

The 2D characteristic patterns from the algorithm are designed with a reason and are generated from the unit cells that have been preselected. The unit cell that is assigned within the block (scaffold) structure is selected based on factors such as the required mechanical and biological properties which have already been predefined. The process defined here assumes that this selection process has been carried out prior to the generation of process-planning instructions. Further details into how these unit cells are selected are described in Sun et al. and the readers are referred to them.

A point that we do not like to disregard is the notable limitations of the process. The approach described here presents an implicit evaluation of the CAD model. The entire scaffold assembly is defined by stand alone CAD models with the unit cells being referenced into the interior of the scaffold. This process will not result in an explicit CAD model and hence cannot be used for visualization purposes or further downstream applications. The user will not be able to visually see the final scaffold in 3D and the "final product" is only seen after fabrication. This may result in costly errors especially when multiple complex unit cells have been used in the interior architecture. However, the application interface helps to avoid such errors by providing tools for the visualization of the 2D patterns that make up the scaffold. One of the major advantages of having a CAD model is being able to use it in downstream FEA analysis packages. Due to the unavailability of an explicit CAD model, scaffolds designed using this approach cannot be subjected to FEA techniques in the conventional manner. However, an indirect method of analysis for effective property characterization can be performed using homogenization-based theories [11].

The process described facilitates the fabrication of heterogeneous models in the structural sense by the inclusion of heterogeneous unit cells. Using a combination of different material architectures within the same structure, true heterogeneity with respect to material and structure can be achieved. By assigning specific material information for each unit cell, the process unlike STL-based methods will carry this information throughout to the point of fabrication. A framework for heterogeneous modeling has been formulated [12,13] and a future extension of the process may be to incorporate material heterogeneity within the scaffold structure. Process-planning instructions generated from the IAD algorithm will then be able to fabricate them using scaffold biofabrication systems [14]. The described process can be further extended to generate process-planning instructions for contour-based RP machines [22]. Although the library of unit cells that can be fabricated with these machines is limited, the authors believe that by proper selection of varied contour patterns and sizes, the desired scaffold properties can be obtained. The process developed for the TheriForm™ machine will further extend its reach to include contour-based RP machines capable of fabricating scaffolds with new biopolymer materials and in micron ranges.

References

1. Robertson DM, Pierre L, Chahal R (1976) Preliminary observations of bone ingrowth into porous materials. J Biomed Mater Res 10:335–344
2. Ryan G, Pandit A, Apatsidis DP (2006) Fabrication methods of porous metals for use in orthopedic applications. Biomaterials 27:2651–2670

3. Langer R, Vacanti JP (1993) Tissue engineering. Science 260:920–926
4. Zeltinger J, Sherwood JK, Graham DA, Mueller R, Griffith LG (2001) Effect of pore size and void fraction on cellular adhesion, proliferation, and matrix deposition. Tissue Eng 7:557–572
5. Chua CK, Leong KF, Cheah CM, Chua SW (2002) Development of a tissue engineering scaffold structure library for rapid prototyping. Part 1: investigation and classification. Int J Adv Manuf Technol 21(4):291–301
6. Nam J, Starly B, Darling A, Sun W. Computer-aided tissue engineering for modeling and design of novel tissue scaffold. In: CAD'04 international conference, Pattaya Beach, Thailand, 24–28 May 2004
7. Lin F, Sun W, Yan Y (2001) A decomposition-accumulation model for layered manufacturing fabrication. Rapid Prototyping J 7(1):24–31
8. Tata K, Fadel G, Bagchi A, Aziz N (1998) Efficient slicing for layered manufacturing. Rapid Prototyping J 4(4):151–167
9. Starly B, Lau A, Sun W, Lau W, Bradbury T (2006) Direct slicing of STEP based NURBS models for solid freeform fabrication. J Comput Aided Des 38(2):115–124
10. Starly B, Darling A, Gomez C, Sun W, Shokoufandeh A, Regli W. Image based Bio-CAD modeling and its application in biomedical and tissue engineering. In: ACM symposium on solid modeling and applications 04, Genova, Italy, 9–11 June 2004
11. Fang Z, Yan C, Sun W, Shokoufandeh A, Regli W (2005) Homogenization of heterogeneous tissue scaffold: a comparison of mechanics, asymptotic homogenization, and finite element approaches. J Appl Bonic Biomech 2(1):17–29
12. Sun W, Hu X (2002) Reasoning boolean operation based CAD modeling for heterogeneous objects. J Comput Aided Des 34:481–488
13. Sun W (2000) Multi-volume CAD modeling for heterogeneous object design and fabrication. J Comput Sci Tech 15(1):27–36
14. Khalil S, Sun W (2007) Biopolymer deposition for freeform fabrication of hydrogel tissue constructs. J Mater Sci Eng C 27(3):469–478
15. Jayanthi Parthasarathy, Binil Starly, Shivakumar Raman. Design of patient specific porous titanium implants for craniofacial applications. In: RAPID 2008 conference & exposition, May 2008, FL, USA
16. Levy HSJ, Chu TW RA, Halloran JW, Feinberg SE (2000) An image based approach for designing and manufacturing craniofacial scaffolds. Int J Oral Maxillofac Surg 29:67–71
17. Hutmacher DW, Schantz T, Zein I, Ng KW, Teoh SH, Tan KC (2001) Mechanical properties and cell cultural response of polycaprolactone scaffolds designed and fabricated via fused deposition modeling. J Biomed Mater Res 55:203–216
18. Kim SS, Utsunomiya H, Koski JA, Wu BM, Cima MJ, Sohn J, Mukai K, Griffith LG, Vacanti JP (1998) Survival and function of hepatocytes on a novel three-dimensional synthetic biodegradable polymer scaffold with an intrinsic network of channels. Ann Surg 228:8–13
19. Sun W, Starly B, Darling A, Gomez C (2004) Computer-aided tissue engineering: application to biomimetic modeling and design of tissue scaffolds. Biotechnol Appl Biochem 39(1):49–58
20. U.S. Scientific Registry for Organ Transplantation and the Organ Procurement and Transplant Network (1990) Annual Report. UNOS, Richmond, VA
21. Yang S, Leong K, Du Z, Chua C (2002) The design of scaffolds for use in tissue engineering. Part 2. Rapid prototyping techniques. Tissue Eng 8(1):1–11
22. Zein I, Hutmacher DW, Tan KC, Teoh SH (2002) Fused deposition modeling of novel scaffold architectures for tissue engineering applications. Biomaterials 23:1169–1185

Chapter 4
Cell Source for Tissue and Organ Printing

Tao Xu, Yuyu Yuan, and James J. Yoo

Abstract Organ printing, a novel approach in tissue engineering, applies computer-driven deposition of cells, growth factors, biomaterials layer-by-layer to create complex 3D tissue or organ constructs. This emerging technology shows great promise in regenerative medicine, because it may help to address current crisis of tissue and organ shortage for transplantation. Organ printing is developing fast, and there are exciting new possibilities in this area. Successful cell and organ printing requires many key elements. Among these, the choice of appropriate cells for printing is vital. This chapter surveys available cell sources for cell and organ printing application and discusses factors that affect cell choice. Special emphasis is put on several important factors, including the proposed printing system and bioprinters, the assembling method, and the target tissues or organs, which need to be considered to select proper cell sources and cell types. In this chapter, characterizations of the selected cells to justify and/or refine the cell selection will also be discussed. Finally, future prospects in this field will be envisioned.

4.1 Introduction

Successful cell and organ printing depends on many key elements, such as bioprinter design [1–3], biopaper (e.g., biomaterial scaffolds) development [4], and bioink (e.g., cells, growth factors) utilization [5]. Among these, the choice of appropriate cells for bioink is vital. Like any other tissue engineering products, organ printing

T. Xu (✉)
Department of Mechanical Engineering, College of Engineering, University of Texas at El Paso, 500 W. University Ave, El Paso, TX, 79968, USA

T. Xu and J.J. Yoo
Wake Forest Institute for Regenerative Medicine, Winston-Salem, NC, 27157, USA
e-mail: txu2@utep.edu

Y.Y. Yuan
Department of Bioengineering, Clemson University, Clemson, SC, 29634, USA

R. Narayan et al. (eds.), *Printed Biomaterials*, Biological and Medical Physics, Biomedical Engineering, DOI 10.1007/978-1-4419-1395-1_4,

constructs are typically composed of cellular and non-cellular components. The source and selection of cell types for the cellular component, therefore, can direct the composition, properties and functions of the printed tissues or organs.

In the design of an organ printing scheme, many important factors need to be considered to select proper cell sources and cell types, and these factors may differ between various applications and between printing systems. For example, when printing the different types of tissue that form an organ, many major cell types are required. Moreover, a number of bioprinters have been developed, and each uses a different printing mechanism. These various printer systems require their own specific cell sources and biomaterial forms for shaping the tissue. In addition, there are diverse approaches to the assembly of tissues and organs, and these each require specialized building blocks for organ printing, including appropriate cell types and biomaterials.

Therefore, choosing appropriate cells and applying them to specific printing systems is becoming critically important in this field. In this section some basic concepts of cell selection for tissue and organ printing will be discussed. After briefly reviewing the available cell sources for tissue and organ printing applications, including cell types and cell presentation forms, factors that affect cell choice will be discussed. Characterizations of the selected cells will also be explored. Finally, future prospects in this field will be discussed.

4.2 Available Cell Sources

There are over 200 different cell types in the human body. Many of these have been established for a variety of tissue engineering applications [6]. The available cells can be classified into two major categories: specialized cells and stem cells. In principle, all of these cells can be printed by bioprinters to fabricate viable tissues or organs. Moreover, the cells can be printed in various forms, and currently three major cell forms are being used in organ printing. These include the individual cell form, the aggregation cell form, and the cell-additive combination form. The available cell sources and forms are categorized and summarized in Fig. 4.1.

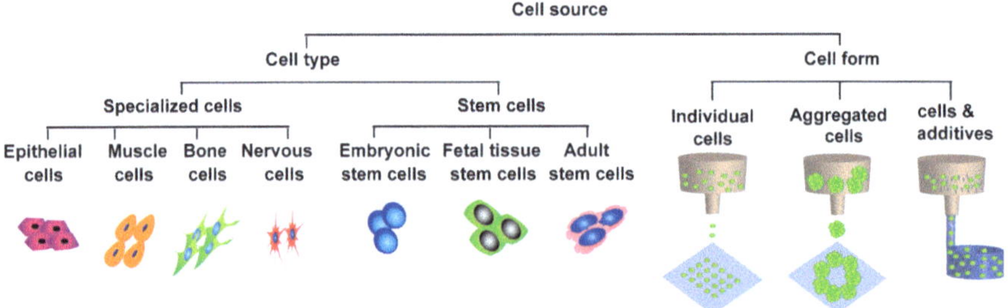

Fig. 4.1 Schematic drawing of cell sources available for organ printing

4.2.1 *Cell Types*

4.2.1.1 Specialized Cells

The most extensively explored cells in tissue and organ printing are specialized cells. This term refers to cells committed to a certain type. Obviously, the phenotype of specialized cells varies from tissue to tissue [7]. Usually specialized cells are obtained by dissociation from a small piece of donor tissue. The source of donor tissue can be heterologous (different species), allogeneic (same species, different individual), or autologous [8]. The preferred cells to use in tissue and organ printing are autologous cells because, although they may cause an initial inflammatory response, autologous cells are histocompatible with the host and avoid rejection. Thus, the deleterious side effects of immunosuppressive medications can be avoided after the tissue printing products are eventually implanted into patients [8].

The first reported specialized cell type used in tissue and organ printing is the embryonic-chick spinal-cord cells [9, 10]. This cell line was printed by a laser guidance direct write system in 1999. Since then, more than 30 different specialized cell types have been used in a range of various printing systems. These cells have been isolated from several major tissue types, ranging from nervous [10–12], cardiac [13], epithelial [6, 14, 15], bone [16–19], and liver [17, 20–22] tissues. In most cases, these cells were originally used to prove the principle of a specific bioprinting system that could print viable materials. The used printing systems include inkjet printing [6, 17, 23], laser guidance direct write [10], laser-induced forward transfer technique [24, 25], extrusion-based printing [21, 26, 27], and electrostatic-based jetting [11, 22, 28]. Table 4.1 summarizes these specialized cell types which have been used in some organ printing practices.

4.2.1.2 Stem Cells

Another promising cell source for tissue and organ printing is stem cells. Although the autologous specialized cells described above are used widely, their source (usually obtained from the diseased organ of the host) and processing method may limit the utility of this type of cell in some applications. For example, a tissue biopsy from a patient with extensive end-stage organ failure may not yield enough normal cells for expansion and transplantation. In other instances, primary autologous human specialized cells cannot be expanded from a required organ. Pancreas, heart, and central nervous tissues present this problem [7]. In these situations, stem cells are envisioned as being an alternative source of cells from which the desired tissue can be derived. Stem cells refer to undifferentiated cells with a high proliferation capacity, the capability of self-renewal, and the potential for multilineage differentiation [8]. Stem cells can be derived from human embryos (human embryonic stem cells) [29], from fetal tissue (amniotic fluid [16], placenta [30]), or from adult sources (bone marrow [31], fat [32], skin [33]).

Table 4.1 Cell sources used for organ printing

Model	Technology	Category	Cell type/cell form	Printed cells
Non-contact (drop-on-demand deposition)	Inkjet based	Thermal inkjet	Specialized cells/ individual	Rat embryonic hippocampal and motoneuron Dog bladder SMC Bovine aortal endothelial cell CHO
			Stem cells/ individual	Human amniotic fluid derived stem cell QCE-6 stem cell
		Piezo inkjet	Specialized cells/i ndividual	Human fibroblast Human osteoblast Bovine chondrocyte Bovine vascular endothelial cell
			Specialized cells/ additives	HeLa cells (alginate)
	Laser based	Laser guidance direct writing	Specialized cells/ individual	Chick embyronic spinal cord neuron Human umbilical-vein endothelial cell Rat hepatocyte Chick embryonic forebrain cell
		Laser-induced forward transfer	Specialized cells/ individual	Human MG-63 osteosarcoma cell Mouse EOMA endothelial cell Rat cardiac cells Rat Schwann and astroglial cell Pig lens epithelial cell Rat B35 neuronal cell
			Stem cells/ individual	Murine embryonal carcinoma P19 cells
	Electrostatic based		Specialized cells/ additive	Mouse neuronal CAD cell Rat hepatocyte
			Stem cells/ individual	Hepatic progenitor cells
Contact (continuous deposition)	Extrusion based	Pressure-controlled head	Specialized cells/ additives	Rat hepotocytes (gelatin gel) Calf articular chondrocytes (alginate gel) Human fibroblasts (F-127 gel) Bovine aortic endothelial cells (collagen gel)
			Specialized cells/ aggregates	CHO aggregates
		Extrusion pneumatic micro-valve	Specialized cells/ additives	Human endothelial and fibroblasts (alginate gel)

Characterization				
Cell viability	Genotypic and phenotypic	Cell property & function	Applications	Refs.
>90%, combined MTS and LDH assay or live/dead assay	MAP2, neuronfilament	Electrophysiology	Cellular pattern	[23, 47]
	αSMC actin	Electrophysiology	3D tissue construct	[19]
	CD31	Contractile properties, intracellular Ca^{2+} mobilization	Blood vessel, vascularized tissue	[14, 19, 48]
			Cellular pattern	[23]
	Transcription factor Oct4	Osteogenic differentiation	Bone-like tissue	[16]
			Cellular pattern	[34]
Most alive, direct observation of cell adhesion and proliferation				[17, 18]
				[17, 18]
				[18]
			Cell arrays	[6]
			3D gel tube and sheet	[49]
89+7% , trypan blue				[9, 10]
			Liver sinusoid-like structure	[15, 20]
			Cellular pattern	[15, 20]
			Cellular pattern	[50, 51]
80–90%, 100% after 1, live/dead assay;Negative to heat shock proteins 60 and 70			Cell array, 3D tissue construct	[24, 38, 42, 52]
			Cell array	[38]
			Cell array	[13]
				[43]
				[43]
	No significant DNA cleavage and apoptosis, TUNEL staining	Extension of neural axons		[44]
	Minimal DNA damage, Alkaline and neutral comet assays	Neurogenic and musculogenic differentiation	Cell array	[25]
Viability similar to non-printed cells, trypan blue or MTT staining				[11, 40, 41]
		Hepatic-specific function	Cell encapsulation	[22]
		Hepatic-specific function	Cell encapsulation	[28]
93% after 1 month, live/dead assay	Human albumin	Normal metabolic activities of hepatocytes	3D hepatic construct	[21]
94+5% ,live/dead assay		Production of cartilage-specific ECM	3D cartilage constructs	[26]
60%, phase contrast			3D cell constructs	[45]
46–86%, phase contrast			Blood vessel	[45]
		Fusion of cell aggregates	Modeling	[36]
85%			Cellular pattern	[53]

The first attempt to experiment with stem cells in tissue and organ printing was conducted with the pluripotent embryonal carcinoma P19 cells [25]. The P19 cell line was used to prove the feasibility of stem cell printing by using a laser forward transfer technique called "Matrix-Assisted Pulsed Laser Evaporation Direct Write" (MAPLE DW) [25]. After printing, the embryonal cells differentiated into neural and muscular lineages under specific stimulus [25]. Another example is the QCE-6 cell line, derived from precardiac mesoderm of the Japanese quail [34]. These cells were printed by using a modified thermal inkjet printer to create specific cellular patterns.

The first cells with direct clinical relevance for tissue and organ printing are human amniotic fluid stem (AFS) cells [16]. We have recently demonstrated that human AFS cells can be printed together with hydrogel scaffolds to build 3D tissue constructs. If printed with osteogenically differentiated human AFS cells, the constructs can form bone-like tissues in immune-deficient mice [16]. It should be pointed out that, although the potential use of stem cells has generated enormous excitement in tissue engineering and regenerative medicine, their use in organ printing is still in its infancy. As advances are made in stem cell biology, other types of stem cells are anticipated to be used in organ printing.

4.2.2 Cell Forms

4.2.2.1 Individual Cells

A basic cell form used in tissue and organ printing is the individual or single cell form. Since the concept of tissue and organ printing launched in the late 1990s, building tissue cell-by-cell with a printer has been the major proposed printing strategy, and individual cells are used as a basic building block for most current organ printing applications. In particular, the individual cells are required in instances where specific bioprinters with smaller printing nozzles are used. For example, in line with the relatively small nozzle diameter, the thermal inkjet or piezo inkjet systems employ individual cells [6, 23]. Moreover, for other specific applications, single cells are also needed. For example, in laser-based printing systems, individual cells instead of cell aggregates are often loaded into the laser based bioprinters to achieve single cell resolution in the cell array [13, 35].

4.2.2.2 Aggregated Cells

Cell aggregates are another important form used for tissue and organ printing. Recently, cell aggregates rather than individual cells have been proposed as an optional building block for printing 3D tissue constructs [3, 5, 36]. In these applications, cell aggregations resulted from genetically transformed cells, like Chinese hamster ovary (CHO) cells, with controlled adhesive properties. These cell aggregates are able to fuse together because of tissue fluidity of cell aggregation. It has been reported that the contiguous CHO aggregates in appropriate gel substrates can fuse or "self-assemble" into cell structures of specified morphology [3, 5]. One advantage of cell aggregates may be the possible acceleration of tissue formation process by means of the quick fusion of the cells [5]. However, to use this cell form, additional steps are needed to modify the cells to make them highly adhesive. This may limit the utility of many cells types, including those not suitable for genetic modification, in organ printing applications that use cell aggregates as "ink."

4.2.2.3 Combination of Cells and Additives

In the two cell forms described above, only cells are used, regardless of whether they are single cells or aggregates (groups). However, in certain printing conditions, the selected cells must be mixed with "bioink" additives, such as hydrogels and other materials, and the subsequent mixture is printed by the bioprinter. The additives used in this form can play a role as either the mechanical supports for building 3D constructs with specific configuration [26, 37]or as the transfer materials for cell delivery [24, 38]. The extrusion-based and electrostatic-based printing systems are two examples of such printing systems. In extrusion-based printing, stimuli-sensitive (thermo-, pH-, and chemo-sensitive) gels are often added to the cell suspension, and the mixture is printed to form 3D constructs in the presence of the chosen stimulus. This induces rapid gelling of the mixture [4]. In electrostatic encapsulation, specific hydrogels, such as alginate, are needed to maintain the form as well as deliver living cells [22, 28].

4.3 Factors to Be Considered for Cell Selection

Ideal cells choices for tissue and organ printing should include these cells types and corresponding forms, which are basic cellular components for target tissues and can be printed with minimal compromise of cellular properties and functions. However, selection of an optimal cell source is not a simple process. Over the past decade a range of bioprinting systems has been developed for different organ printing applications. However, not all printing systems can use the same cell types, so therefore cell selection is difficult to standardize and simplify the cell selection process. Moreover, the selection process is also complicated by the fact that various assembling strategies have been proposed for the same organ. These and other important factors affect the choice of cells used in organ printing practices. We will focus on some of these key factors to be considered before actual printing.

4.3.1 Target Tissues or Organs

The first aspect to be considered is the target tissue or organ to be fabricated by the organ printing approach. Most tissues or organs contain multiple cell types, and each cell type plays a distinct role in proper organ function. Moreover, the cell types involved vary from tissue to tissue and from organ to organ. Therefore, the cell types used for printing must be chosen very carefully, and they must match the composition of the organ as it would appear in vivo. For example, when attempting to print skin tissue, the main epidermal cell types such as keratinocytes, melanocytes, Merkel cells, Langerhans cells, and others in any combination may be needed [39]. Similarly, to assemble bone tissues, the main bone cell types, such as osteoblasts, osteoclasts, osteocytes, and any combination of these, may be needed [39]. In addition, to print a specific tissue or organ in which primary autologous cells cannot be expanded, such as pancreas or central nervous system, suitable stem cells instead of specialized mature cells might be needed for printing those tissues or organs.

To arrange these different types of cells in a way that will allow them to regenerate the target tissues or organs, the proposed printing system must have the ability to simultaneously print different cell types. Recently, Barron and co-workers have demonstrated that human MG-63 osteosarcoma cells and mouse EOMA endothelial cells can be printed onto different locations on a slide by using laser printing (MAPLE DW) [38]. More recently, we have shown that more than three different cell types can be printed into a "pie"-type multicell configuration by using a modified thermal inkjet printer. Each of the cell types demonstrated viability and normal function both in vitro and in vivo after printing [19].

4.3.2 Printing Systems and Bioprinters

Over ten different organ printing systems have been developed to date (Fig. 4.1). Most of them are "borrowed" or adapted from industry designs, especially rapid prototyping [1]. These printing systems include the very first printing system, laser guidance direct writing [9], and the most popular method, inkjet printing [23]. Recently, a new type of cell printing, electro-hydrodynamic jetting, was developed [40]. In these different printing systems, physical, chemical, and biological processes occur distinctly and may have different influences on the properties and functions of the selected cells. Therefore, when selecting a proper cell source, another important factor to be considered is whether the selected cells or cell forms will fit the needs of the printing system to be used. For example, in the thermal- and piezo-based inkjet printing systems, the printhead's nozzle size is limited to 30–100 μm, and cells larger than that cannot pass through these nozzles [6, 12]. Therefore, cell types with a smaller size are optimal for these systems.

The electro-hydrodynamic jetting system is another example. Most of the current electrostatic-based systems like electro-spinning and electro-spraying, fail to provide a stable jet model for precise delivery of living cells to specific target locations as inkjet printing does. To overcome this, a novel coaxial needle design has recently been introduced. Here, a concentrated cell suspension flows through the inner needle and a medical grade poly(dimethylsiloxane) (PDMS) medium with high viscosity and low electrical conductivity flows through the outer needle [41].

With this modification, the electro-hydrodynamic jetting system can print cells in a controlled manner, and can form products such as microthread structures [41]. In addition, in the extrusion systems, to ensure that more cells survive the harsh extrusion process during printing, the cells are often mixed with the biocompatible hydrogels or polymers for printing.

4.3.3 Assembling Approaches

The basic concept of "building" tissue or organs layer-by-layer from individual components has not changed from the first organ printing experiments. To achieve this goal, appropriate assembling approaches are needed. It is generally accepted that there are three different approaches for assembling organs or tissues. These approaches have been termed "structural," [35] "conformal," [35] and "aggregation" [3, 5] organ printing. Structural printing requires that the same tool print the scaffolding, cells, and biomolecules simultaneously or sequentially [35]. Conformal printing is a hybrid approach that prints cells and biomolecules on top of thin layers of prefabricated scaffolding [35]. Aggregation organ printing uses cell aggregates as the printing units or assembling blocks, in an attempt to prompt tissue formation by means of specific fluidic properties [5]. In each of these strategies, specific cells and cell forms are needed. In conformal printing, individual cells are usually needed. In structural printing, a mixture of cells and an additive are used to create simultaneous or serial printing of cells, scaffolds, and biomolecules. Aggregation printing requires that cells be prepared through genetic modification with adhesion molecules prior to printing, so that the cells will stick together to form the desired shape.

4.4 Characterization of Selected Cells

To justify and/or refine cell selection, rigorous characterization of the selected cells is required in tissue and organ printing. Moreover, because most of the current tissue and organ printing systems involve complex and harsh physical and chemical processes, and these processes might significantly affect the properties and fate of the printed cells, it is necessary to evaluate the properties and function of the cells after being printed as well. Important aspects that need to be tested after organ printing include cell viability and proliferative capacity, genotype and phenotype, and cellular function.

4.4.1 Viability and Proliferation

At the cellular level of tissue and organ printing, the first important concern is whether or not the cells remain viable and able to proliferate after they have undergone the printing process. Most current bioprinters were adapted from industrial designs and methods; for example, rapid prototyping techniques. These techniques involve the use of hostile solvents and toxic buffers, and/or they involve harsh mechanical processes. Therefore, when research into printing techniques was begun, it was critical to determine whether the harsh physical and chemical processes required for printing would damage cells and reduce viability and normal function.

The most common methods used to evaluate viability in organ printing include the live/dead assay [19, 26, 38, 42], trypan blue staining [10, 43], TUNEL staining for apoptosis [44], and direct observation of cell adhesion and proliferation [6, 17, 45]. MTT or MTS assays are used for measurement of cell proliferation; this assay measures the metabolic activity of the printed cells [19] using a colorimetric assay, and it is assumed that increased metabolic activity (increased color) in a sample is related to a higher number of cells present.

After stringent evaluation of biocompatibility used the above-mentioned assays, most of the printing systems demonstrated satisfactory results in terms of survival and viability, which ranged from 46 to 100% [35, 45] depending on the printing system. Moreover, no significant slowing of cell growth after printing was observed in these systems [19, 42].

There are several generally accepted explanations for these high viabilities and subsequent normal proliferation. These include the low power used in the system, particularly with laser guidance direct writing [9, 10], and the short time that cells are under stress, which is typically on the order of microseconds for most printing experiments, such as inkjet printing [23, 46] and laser printing [13, 23, 42, 46]. The hostile physical and mechanical effects, like heat, occurred in the printing system, but their effects on the cells to be printed were minimized by short exposure times. Therefore, the cells would survive in most tissue and organ printing systems.

Another possible mechanism for high cell survival rates might lie in the morphological changes of the cells to be printed. We have termed this the "cell ball" theory. To obtain a cell suspension for printing, trypsin or other ECM proteases are often used to dissociate or remove the cells from cultured substrates or donor tissues. Such enzymatic treatment creates a single-cell suspension. When cells are singular, rather than organized in a tissue, their morphologies change from a fully "spread-out" status into a "ball-like" form. It is these "ball-like" cells that are loaded into the printing system for printing.

With these significant morphological changes, the surface area of the cells diminishes greatly, meaning there is less surface area available for interaction with outer stimuli. This "cell ball" theory, combined with the above explanations for cell viability, could also explain, in part, how basic properties and functions of printed cells are maintained, as described below.

4.4.2 Cell Genotype and Phenotype

Each cell has its distinct genotype and phenotype, and these features can easily change in reaction to environmental stimuli. Therefore, another concern during printing cells is whether harsh printing processes could adversely affect genetic information (DNA) as well as expression of the normal phenotype of the printed cells. To detect potential DNA damage caused by the physical trauma incurred in the printing process, especially the UV laser light used in the MAPLE DW process, the comet assay is usually used. For example, the alkaline and neutral comet assays were performed to determine whether the MAPLE DW process induces single-brand or double-strand breaks in the DNA of the transferred cells, respectively [25]. The results showed no observable DNA damage from potential cell–laser interaction [25].

The staining of heat shock proteins in printed cells from a laser printing system is another example [42]. To experimentally determine the amount of damaging heat and/or shear stress that printed human osteosarcoma cells experience during the laser printing process, antibodies specific to heat shock proteins 60 and 70 (anti-HSP60/HSP70 have been used) [42]. Minimal expression of stress-induced proteins was found, which suggests that the printed cells were exposed to little environmental damage [42].

To evaluate the phenotypes of the printed cells, immunohistochemistry is usually employed. Cell specific antibodies are used to determine whether printed cells maintain their phenotypes. For example, the neuronal markers microtubule-associated protein (MAP)-2 and neurofilament NF150 were used to analyze hippocampal and cortical cells after they passed through the firing nozzles of a thermal inkjet printer [12, 34]. Printed dog bladder smooth muscle cells (SMC), bovine aortal endothelial cells, and human amniotic fluid stem (AFS) cells were characterized with αSMC actin, CD31, and the transcription factor Oct4, respectively [19].

4.4.3 Functional Analyses

A requirement for building functional tissues or organs with organ printing is that the printed cells should display normal cell properties and functions. Therefore, it is necessary to evaluate whether the cells can maintain such important features after being printed. The methods used for functional analysis vary from cell type to cell type. Electrophysiological analysis has been used to evaluate functions of printed neuronal cells [12, 34] and muscle cells [19]. For example, after being printed with an inkjet system, hippocampal and cortical neurons demonstrated normal sodium ion (Na^+) currents and action potential firing in patch clamp experiments [12, 34]. Sodium channel function and the ability to induce action potentials are two of the most characteristic features of neurons. Similarly, bladder smooth muscle exhibited similar potassium ion (K^+) channel properties to control cells post-printing [19]. Moreover, functional evaluation of epithelial cells was performed with analysis of intracellular calcium (Ca^{2+}) levels in response to pharmacological stimuli, which is one of the common methods to test basic epithelial cell function [19].

With many types of stem cells being enlisted into organ printing applications, effective methods to evaluate the function of printed stem cells are needed. One of the most important properties of stem cells is their pluripotency, or their potential to differentiate into many different cell types; therefore, whether stem cells can maintain such an important capability during the printing process is becoming a central issue for stem cell printing. To this end, after being printed, stem cells have been tested with different lineage differentiation protocols. Barron et al. reported that pluripotent human embryonal carcinoma P19 cells could differentiate into neural and muscular lineages under specific stimulus after being printed with a laser forward transfer printer [25]. We have recently demonstrated that AFS cells could be differentiated into osteogenic cells in vitro and in vivo [16, 19].

4.5 Conclusions

An initial and critical step in tissue and organ printing, the choice of cells will determine the properties and fate of the final products. Selection of specific cell sources for certain organ printing applications is not a simple process, and it depends on many factors, such as the proposed printing system and bioprinters, the assembling method, and the target tissues or organs. Moreover, to justify and/or refine the cell selection, characterizations of the printed cells are required to verify basic cellular properties and functions to make certain the process does not damage the cells.

Currently, the race to develop advanced tissue and organ printing systems and apply organ printing technology to clinical problems is underway. There will be many exciting advancements and new possibilities in this area. With further development of regenerative medicine and organ printing, there will a growing number of cell sources available for organ printing application. Moreover, with the development of more sophisticated printing systems and the introduction of new biomaterials, the process of cell selection will be adapted to these new conditions. In conclusion, utilization of stem cells is envisioned as the future of tissue and organ printing.

References

1. Boland T, Xu T, Damon B et al (2006) Biotechnol J 1(9):910
2. Mironov V (2003) Expert Opin Biol Ther 3(5):701
3. Mironov V, Boland T, Trusk T et al (2003) Trends Biotechnol 21(4):157
4. Fedorovich NE, Alblas J, de Wijn JR, et al (2007) Tissue Eng
5. Jakab K, Neagu A, Mironov V et al (2004) Proc Natl Acad Sci U S A 101(9):2864
6. Nakamura M, Kobayashi A, Takagi F et al (2005) Tissue Eng 11(11–12):1658
7. Polak JM, Bishop AE (2006) Ann N Y Acad Sci 1068:352
8. Atala A (2004) Rejuvenation Res 7(1):15
9. Odde DJ, Renn MJ (1999) Trends Biotechnol 17(10):385
10. Odde DJ, Renn MJ (2000) Biotechnol Bioeng 67(3):312

11. Eagles PA, Qureshi AN, Jayasinghe SN (2006) Biochem J 394(Pt 2):375
12. Xu T, Gregory CA, Molnar P et al (2006) Biomaterials 27(19):3580
13. Barron JA, Ringeisen BR, Kim H et al (2004) Thin Solid Films 453–454:383–387
14. Boland T, Tao X, Damon BJ et al (2007) Mater Sci Eng Biomim Supramol Syst 27(3):372
15. Nahmias Y, Schwartz RE, Verfaillie CM et al (2005) Biotechnol Bioeng 92(2):129
16. De Coppi P, Bartsch G Jr, Siddiqui MM et al (2007) Nat Biotechnol 25(1):100
17. Saunders R, Bosworth L, Gough J et al (2004) Eur Cell Mater 7(Suppl. 1):84
18. Saunders R, Derby B, Gough J et al (2005) Mater Res Soc Symp Proc Mater Res Soc 845:57
19. Xu T, Zhao W, Atala A, et al Presented at the Digital Fabrication 2006 (unpublished)
20. Nahmias Y, Odde DJ (2006) Nat Protoc 1(5):2288
21. Yan Y, Wang X, Pan Y et al (2005) Biomaterials 26(29):5864
22. Zhou Y, Sun T, Chan M et al (2005) J Biotechnol 117(1):99
23. Xu T, Jin J, Gregory C et al (2005) Biomaterials 26(1):93
24. Barron JA, Spargo BJ, Ringeisen BR (2004) Appl Phys A Mater Sci Process 79:1027–1030
25. Ringeisen BR, Kim H, Barron JA et al (2004) Tissue Eng 10(3–4):483
26. Cohen DL, Malone E, Lipson H et al (2006) Tissue Eng 12(5):1325
27. Sun W, Darling A, Starly B et al (2004) Biotechnol Appl Biochem 39:29
28. Chandrasekaran P, Seagle C, Rice L et al (2006) Tissue Eng 12(7):2001
29. Atala A (2005) J Urol 174(6):2085
30. Delo DM, De Coppi P, Bartsch G Jr et al (2006) Methods Enzymol 419:426
31. Lajtha LG, Porteous DD (1965) Minerva Nucl 9(4):203
32. Gimble JM, Katz AJ, Bunnell BA (2007) Circ Res 100(9):1249
33. Bartsch G, Yoo JJ, De Coppi P et al (2005) Stem Cells Dev 14(3):337
34. Xu T, Gregory C, Molnar P, et al Presented at the Materials Research Society Proceeding, 2005 (unpublished)
35. Ringeisen BR, Othon CM, Barron JA et al (2006) Biotechnol J 1(9):930
36. Neagu A, Jakab K, Jamison R, et al (2005) Phys Rev Lett 95(17): 178104
37. Yang SF, Leong KF, Du ZH et al (2002) Tissue Eng 8(1):1
38. Barron JA, Wu P, Ladouceur HD et al (2004) Biomed Microdevices 6(2):139
39. Marieb EN (2000) Human anatomy & physiology, 5 ed. Benjamin/Cummings, Menlo Park, CA
40. Jayasinghe SN, Qureshi AN, Eagles PA (2006) Small 2(2):216
41. Townsend-Nicholson A, Jayasinghe SN (2006) Biomacromolecules 7(12):3364
42. Barron JA, Krizman DB, Ringeisen BR (2005) Ann Biomed Eng 33(2):121
43. Hopp B, Smausz T, Kresz N et al (2005) Tissue Eng 11(11–12):1817
44. Patz TM, Doraiswamy A, Narayan RJ et al (2006) J Biomed Mater Res B Appl Biomater 78(1):124
45. Smith CM, Stone AL, Parkhill RL et al (2004) Tissue Eng 10(9–10):1566
46. Xu T, Petridou S, Lee EH et al (2004) Biotechnol Bioeng 85(1):29
47. Xu T, Gregory C, Molnar P, et al (2006) Biomaterials 27(19): 3580
48. Kesari P, Xu T, Boland T Presented at the Materials Research Society Proceeding, 2005 (unpublished)
49. Nakamura M, Nishiyama Y, Henmi C, et al Presented at the Digital Fabrication 2006 (unpublished)
50. Pirlo RK, Dean DM, Knapp DR et al (2006) Biotechnol J 1(9):1007
51. Bakken DE, Narasimhan SV, Burg KJL et al (2005) Macromol Symp 227:335
52. Doraiswamy A, Narayan RJ, Harris ML et al (2007) J Biomed Mater Res A 80(3):635
53. Khalil S, Nam J, Sun W (2005) Rapid Prototyping J 11(1):9

Chapter 5
Direct-Writing of Biomedia for Drug Delivery and Tissue Regeneration

Salil Desai and Benjamin Harrison

Abstract This chapter presents direct-write methods for precisely depositing biomedia for drug delivery and tissue engineering applications. Specifically, different inkjet methods, their operational modes and drop generation dynamics are detailed. Some of the unique challenges for inkjetting biopolymers and the control of their rheological properties are highlighted. The manufacturing of drug delivery microcapsules with controlled release kinetics based on variations in inkjetting and fluid properties is discussed. Finally, the inkjetting of biomedia including stem cells and growth factors into a complex 3D construct for tissue regeneration is elaborated.

5.1 Introduction

With greater understanding of how biological systems work, comes the potential for unlocking numerous benefits to humanity. Some of this success can be attributed to the better development of tools and materials for probing and interacting with living systems. Development of custom medicines and tissue-engineered organs are just a few examples where understanding not only biology but also control of materials have mutually benefited each other. For these fields to continue to advance will require greater understanding how to assemble not only biologically relevant materials but also living systems such as cells to complex structures.

The ability to precisely deposit biomedia, which include nucleic acids, proteins, and oligonucleotides, is also the foundation for developing various research and diagnostic applications [1–3]. Currently there are several approaches to depositing biomedia that have relied on either top-down [4, 5] and bottom-up approaches [6, 7]. One promising approach is direct-writing, which defined as the selective placement

S. Desai (✉)
North Carolina A&T State University, Greensboro, NC 27411, USA

B. Harrison
Wake Forest Institute for Regenerative Medicine, Winston-Salem, NC 27157, USA

R. Narayan et al. (eds.), *Printed Biomaterials*, Biological and Medical Physics, Biomedical Engineering, DOI 10.1007/978-1-4419-1395-1_5,

or deposition of biomedia at target-specific sites. Based on the type of direct-writing technique utilized it is possible to jet, scribe, or trace different biopolymers on substrates for different biomedical applications. One significant advantage of direct-writing methods over conventional biomanufacturing techniques is as they prevent contamination of the target site when delivering biopolymers and relevant nutrients [8].

Two areas where direct-writing methods are gaining traction is for advanced drug delivery and tissue engineering. These fields can greatly benefit from the precision control that direct-writing methods can provide.

In recent years, biopolymer microcapsules are used as alternate drug delivery carriers with controlled-release of encapsulated drugs in optimum dosage for extended periods of time. These novel drug delivery systems increase the efficacy of highly toxic drugs by localizing the impact of the drug and minimizing patient complications which is common during systemic intake [9].

Tissue engineering involves the growing of relevant cell(s) in vitro into a three-dimensional (3D) organ or tissue construct. Biopolymer tissue scaffolds provide a biodegradable architecture to seed growth factors, cells and relevant biomaterials for augmented recovery and regeneration of host tissue.

The focus of this chapter is the applicability of inkjet-based direct-write approach to drug delivery carriers and tissue engineering constructs. First, the basic differences in inkjetting approaches will be explained. Also, some of the unique challenges that inkjetting biopolymers create will be highlighted. Next, examples how inkjetting can be applied to drug delivery systems will be illustrated. Finally, applying inkjetting techniques to tissue engineering will be discussed.

5.2 Inkjet Printing

Simplistically, inkjet printing involves the ejection of microdroplets of precise dimensions from a small aperture directly to a specified position on the substrate to create an image. These microdroplets can be loaded with different compositions to suit the application intent [10–14]. While this technique is well-established for printing colored inks to form images and text, over the past decade inkjet printing technology has been increasingly applied toward biomedical applications [15–19]. The attraction is that microdrops can act as carriers for biological compounds and cell lines with precise metering capability for the intended payload.

Inkjet printing can be divided into two major classes which relate to their modes of operation: the continuous inkjet (CIJ) and drop-on-demand technique [20]. In the continuous inkjet mode, fluid is supplied at high pressure within a nozzle assembly to generate fluid jet. A piezoelectric member within the nozzle assembly vibrates at high frequency perturbing the fluid jet. The propagation of waves forms standing nodes within the jet resulting in consistent droplets in size and spacing [21]. An advantage of the CIJ method is the generation of microdroplets at high rates; however, it has limited placement accuracy [22]. The second class of inkjet printing is the Drop-on-demand (DOD) inkjet system which uses acoustic pulse to eject microdroplets from the nozzle orifice at ambient fluid pressure [23].

Based on the source of pulse, (DOD) inkjet technology is further classified as thermal and piezoelectric driven printing technology. In the thermal DOD printer, a microdrop is ejected by a vapor bubble formed by instantaneous heating of the fluid. Commercial thermal inkjet printers have been used to print biomolecules with minimal loss of bioactivity for cell configurations [24], protein arrays [25] and DNA chips [26]. The use of thermal inkjet printers may be limited for printing animal and human cells as they are generally sensitive to heat [27]. The second type of DOD inkjet printing is the piezoelectrically (PZT) driven inkjet printer which uses the deformation of a piezoelectric membrane to produce an acoustic waveform for fluid ejection. Due to the absence of heat fluxes this type of printing approach may be more suitable for polymer systems that are impregnated with sensitive biomolecules.

Figure 5.1 illustrates a schematic of the drop-on-demand inkjet system manufactured by MicroFab Technologies Inc. Plano, TX. The system consists of wave form generator and amplifier (JetDrive III), a pneumatics console, optics system and piezoelectric nozzle assembly.

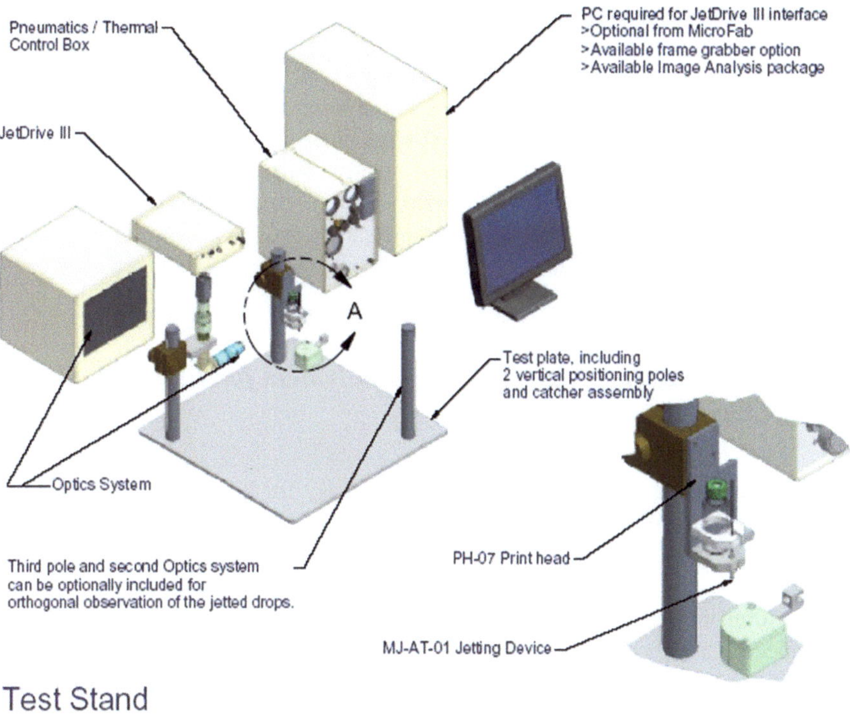

Fig. 5.1 Schematic of the MicroFab Drop-on-Demand Inkjet System (Courtesy: MicroFab Technologies Inc., Plano, TX)

The piezoelectric nozzle assembly is designed to dispense single drops of solvents, water-based fluids and inks. With adequate fluid preparation and device maintenance, the jetting device provides reliable delivery of fluid microdrops. Fluid with viscosities less than 40 centipoise (cP) and surface tensions in the range of 20–70 dyn/cm can be successfully jetted. Viscosities of high molecular weight biopolymers can be lowered by heating the nozzle assembly. The formation of consistent microdroplets is dependent on the wetting behavior of the fluid with the nozzle orifice. Meniscus stabilization is performed to retract hydrophilic fluid and protrude hydrophobic fluid from the nozzle orifice. Depending on the viscosity, surface tension, and contact angle of the fluid with the nozzle tip a positive or negative holding pressure is used to perform meniscus stabilization.

To generate microdrops, an acoustic pulse is sent to the piezoelectric (PZT) member that encapsulates a glass capillary that holds the fluid. By applying a voltage to the PZT actuator, the cross-section (acoustic impedance) of the capillary tube is reduced/increased producing pressure variations of the fluid enclosed in the tube. These pressure variations cause the fluid to disseminate in the glass tube toward the orifice forming a microdrop. Drops of various sizes and volumes can be produced by adjusting the voltage and waveform parameters. Figure 5.2 shows the different phases of typical voltage waveform for jetting Newtonian fluids. The waveform parameters include t_{rise} = initial rise time; t_{dwell} = time at high voltage (V_1); t_{fall} = transition time from high voltage to low voltage; t_{echo} = time at low voltage (V_2); t_{frise} = final rise time.

The rise and fall times in most cases are around 3–5 μs, and the dwell times (durations of the positive and negative voltage pulse plateaus) are normally in the range of 15–50 μs. The falling edge of the positive pulse excursion determines the release time of the drop from the nozzle. A high speed charge-coupled device (CCD) camera with a microscopic zoom lens can be used to observe the formation and trajectories of droplets in flight from the tip of the nozzle.

5.2.1 Rheological Properties of Biopolymers for Inkjet Applications

A critical aspect of using inkjet printing technology for bioprinting is the precise control of fluid properties. Specifically, viscosity and surface tension are the two important properties that determine the ability to jet consistent monodisperse droplets. As the viscosity and surface tension of the polymer increases the resistance to flow within the nozzle increases thereby requiring higher voltages to drive the piezoelectric member within the inkjet nozzle. In general, biopolymer solutions are diluted to lower viscosities to make them amenable to jetting. Typical fluid viscosities for inkjet printing via the DOD inkjet technology lie around 20 cP to minimize viscous dissipation of the kinetic energy [28].

However, very low viscosity biopolymer solutions tend to have a shear thinning behavior leading to excessive tail formations behind the parent microdrop.

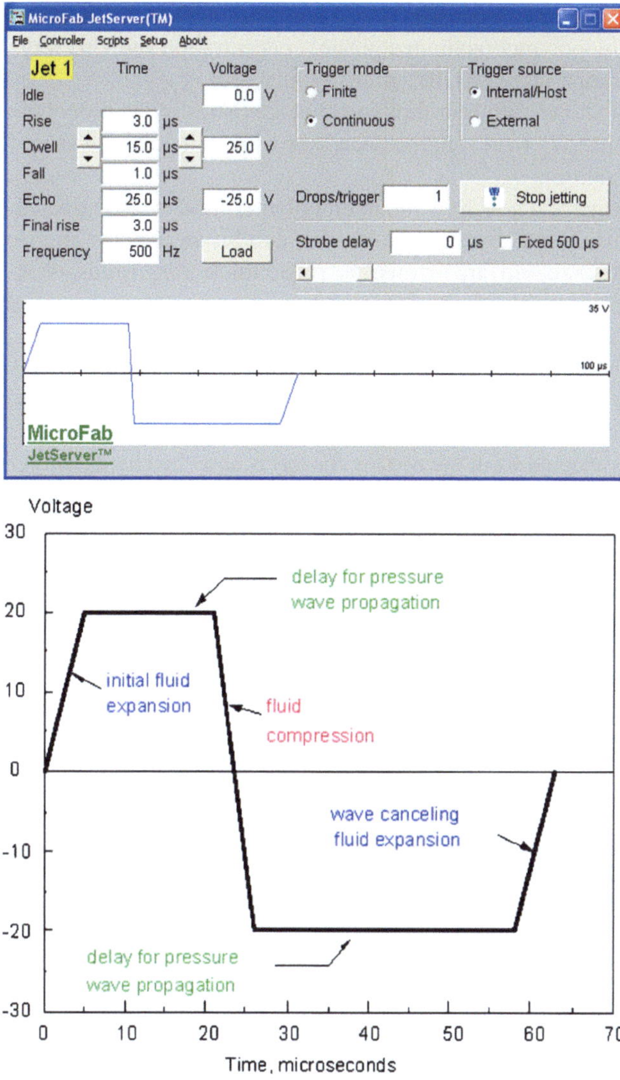

Fig. 5.2 Piezoelectric member excitation waveform (Courtesy: MicroFab Technologies Inc., Plano, TX)

Satellites droplets are extraneous droplets far smaller than the parent microdrop. Satellite drops are formed due to disintegration of long tails that hampers the accuracy of printed features.

To optimize input parameters to achieve monodisperse biopolymer droplets it is important to understand its rheological behavior. Generally, fluid solutions are filtered before ejecting them from the nozzle to prevent clogging problems. The viscosity of biopolymer solution is a function of the temperature, pressure, shear rate, and the degree of concentration [29]. Figure 5.3 depicts a nonlinear relationship between the biopolymer (sodium alginate) concentration and viscosity. As the polymer concentration increases, the viscosity increases.

Figure 5.4 shows the shear thinning characteristics of the biopolymer fluid. The viscosity of the 2% (w/v) solution decreases with an increase in shear rate 22.98°C (room temperature).

The viscosity of a non-Newtonian biopolymer fluid is also affected by temperature variations. Non-Newtonian fluids that exhibit an increase in viscosity as the shear rate is increased are known as shear thickening fluids. When heated to 37.25°C, the viscosity of the 2% (w/v) solution increases as the shear rate is increased as shown in Fig. 5.5.

The non-Newtonian fluid property of a biopolymer solution affects the manner in which the droplets are formed after exiting the nozzle. As the viscosity of the fluid increases the tail length formed behind the parent drop increases. As the drop is completely ejected from the nozzle, the tail is absorbed by the drop thus increasing its volume and diameter. Figure 5.6 illustrates that the length of the tail following the lead drop increased as the polymer concentration is increased. Correspondingly, Fig. 5.7 shows the resultant increase in droplet diameter as the polymer concentration is increased. These images are captured just before the tail broke away from the nozzle tip.

5.2.2 *Drug Delivery Carriers*

Drug delivery is a critical factor in the use of high potency therapeutics. Pharmaceuticals have been administered via intravenous routes, topical treatments,

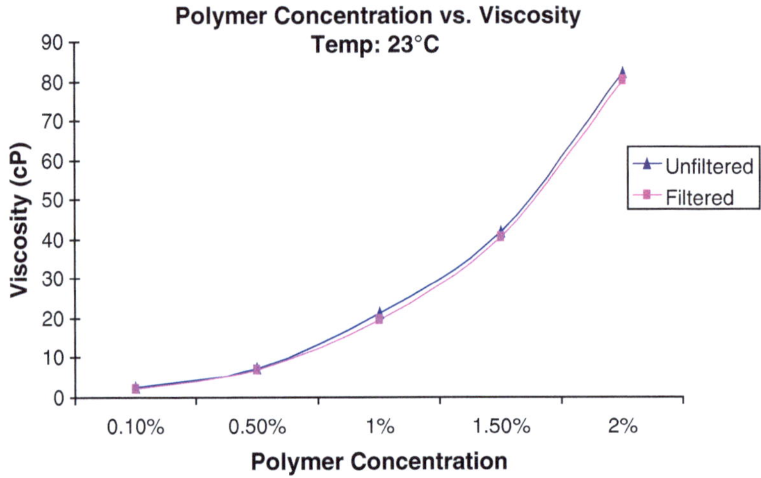

Fig. 5.3 Viscosity vs. Polymer Concentration (filtered vs. unfiltered)

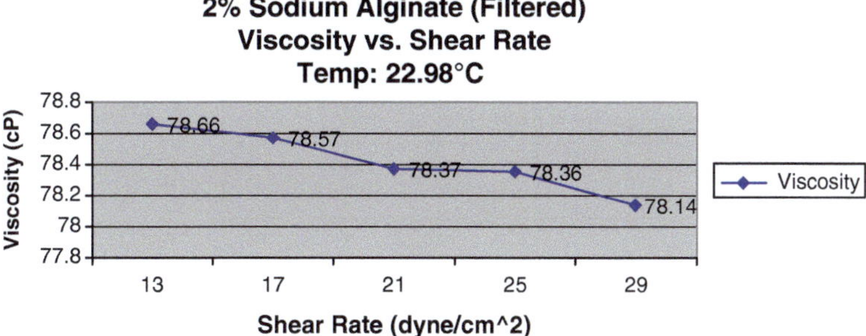

Fig. 5.4 Shear thinning phenomenon for 2% (w/v) solution

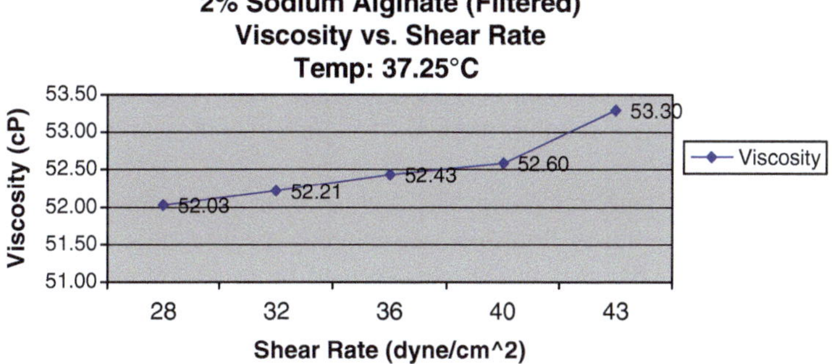

Fig. 5.5 Shear thickening phenomenon of 2% (w/v) solution

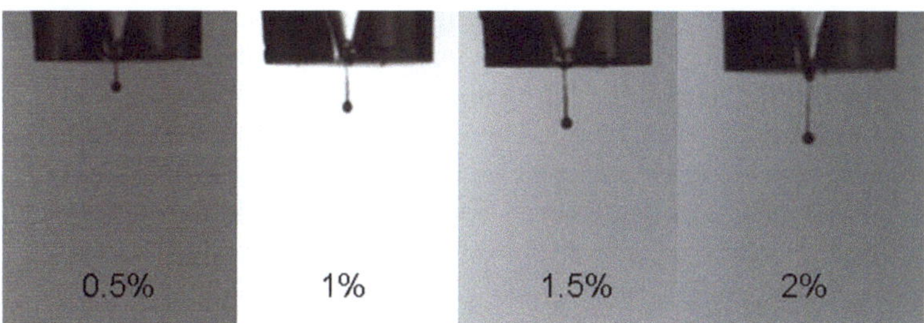

Fig. 5.6 Length of tail filament per sodium alginate concentration (%w/v) (Courtesy: Moore A., NC A&T SU)

and more commonly, orally via liquids, capsules, and tablets. However systemic side effects associated with large doses of medications limit their efficacy in treatments. Another concern is lower bioavailability of drugs due to limited gastrointestinal assimilation efficiency. The concept of localizing drug delivery via alternate

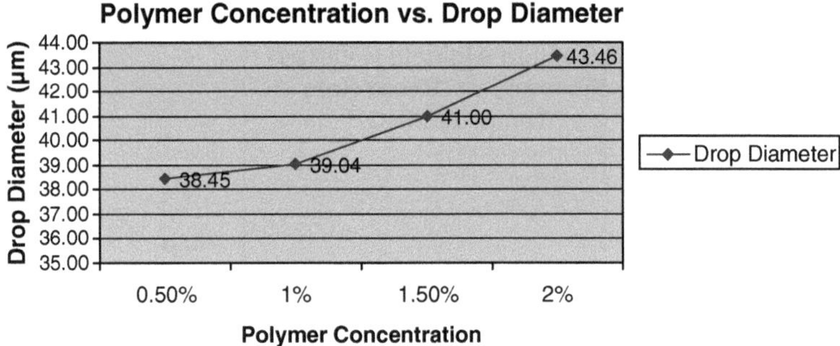

Fig. 5.7 Effect of polymer concentration on drop diameter

administration routes has potential for maximizing patient safety at reduced treatment costs. Microdevice and encapsulation technologies have been proposed for controlled drug delivery in vivo [1, 30]. There are, however, two significant barriers in terms of cost and reliability which have hindered their uptake in clinical practice. Primarily, methods that are both cost-effective and have adequate quality control are lacking. Secondly, a reliable method of controlling the varying rate of drug release is needed.

The use of biopolymer microcapsules such as hydrogels is an effective method for delivering high potency drug loads to ailment specific sites within the body. Hydrogels are water-based colloidal gels formed by cross-linking of hydrophilic polymer chains [31]. Hydrogels can retain upto 90% of water of their total weight depending on the nature of polymer and cross-linking density [32]. Hydrogels possess desirable properties for biomedical applications that include biodegradability, adaptable swelling, and tissue-mimicking behavior. Current microcapsule fabrication methods include electrostatic ionotropic gelation, emulsion foam freeze-drying, filter precipitation that form hydrogel drug delivery carriers with a wide range of size distributions. However, the variations in drug delivery carrier sizes can adversely affect the reliability of drug dosage being administered.

Direct-write inkjet technology is an innovative method for manufacturing of biopolymer microcapsules with precise size distributions. Using this technology manufacturing rates can be achieved between 1,000 to 5,000 microcapsules per second. Inkjet technology can be scaled up (MHz – 10^6 ranges) for commercial purposes using a parallel array of multiple print heads.

Several natural and synthetic polymeric systems have been explored for the controlled release of therapeutic agents. Most prominently among these include alginate-based biopolymers that facilitate the encapsulation and delivery of different biological agents [33]. Sodium alginate is commonly used in the food, drinks, pharmaceutical, textile, and bioengineering industries. Alginate is nontoxic and completely biodegradable [34] within the human body, which makes it an excellent biomaterial for the fabrication of tissue scaffolds.

Sodium alginate ($NaC_6H_7O_6$) is a natural polysaccharide extracted from seaweed that contains two uronic acids, β-D-mannuronic acid (M) and α-L-guluronic acid

(G) [35]. Soluble sodium alginate can be cross-linked by polyvalent cations leading to hydrogel formation [36]. During cross-linking sodium ions are substituted by calcium ions [37], as in the following reaction:

$$Na_2(Alginate) + Ca^{2+} \rightarrow Ca(Alginate) + 2Na^{+}$$

The calcium ions bind to multiple carboxyl groups simultaneously thereby bridging polymer chains to form the hydrogel. A typical calcium alginate hydrogel can be prepared by depositing sodium alginate biopolymer solution into calcium chloride solution to initiate cross linking of polymeric chains and vice versa. During the cross-linking process, other materials present in the environment may become trapped within the hydrogel matrix. For example, calcium alginate-based microcapsules have been used for encapsulating biomaterials including cells [38] and DNA [39] with retention of their biological activity.

This simple process of combining and cross-linkable biopolymers can be controlled in a precise manner using the inkjet process. Figure 5.8 [40] shows microcapsules and scaffolds of calcium alginate hydrogels manufactured using drop-on-demand inkjet printing within aqueous media. The size (diameter) of the microcapsules is dependent on the precision of the microdrop being generated from the nozzle head. One of the main benefits of using inkjet technology is its consistency in manufacturing tight distribution microcapsules. Figure 5.9 shows the scanning electron microscope (SEM) image of a microcapsule (40 μm in diameter).

Depending on the concentrations of biopolymer and calcium chloride solutions, hydrogels of varying porosity and wall thickness can be formed. Figure 5.10 shows the formation of calcium alginate microcapsules when 1.5% (w/v) sodium alginate biopolymer is deposited into (a) 0.1M and (b) 0.25M calcium chloride solutions, respectively. An increase in calcium chloride concentration results in higher cross-linking of polymeric chains forming calcium alginate microcapsules with denser walls.

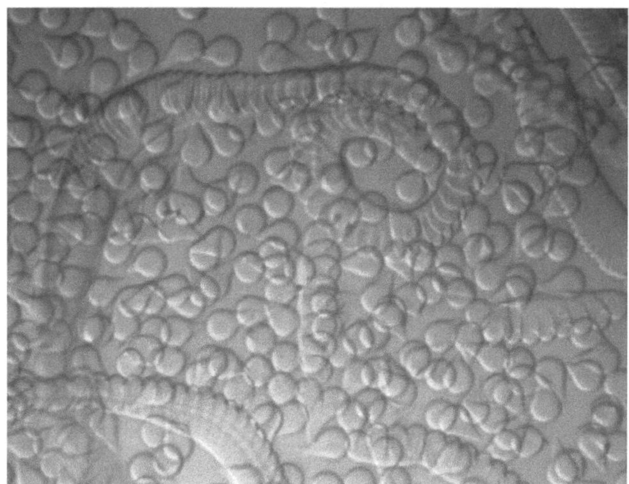

Fig. 5.8 Calcium alginate microcapsules and scaffold constructs

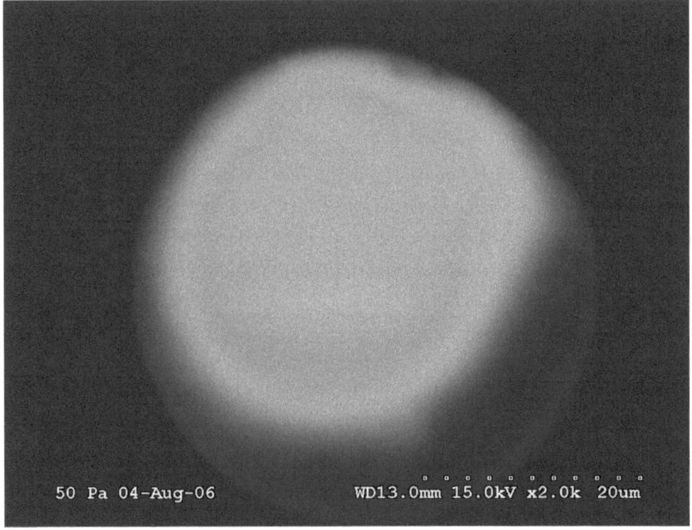

Fig. 5.9 Scanning electron microscopy image of calcium alginate microcapsule (40 μm in diameter)

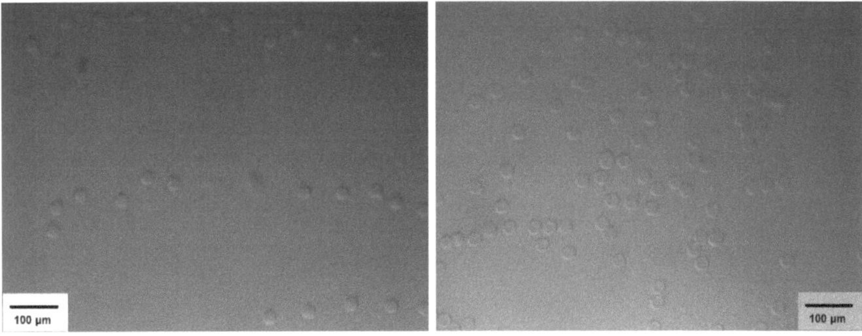

Fig. 5.10 1.5% Ca–Alg microcapsules in (**a**) 0.1M $CaCl_2$; (**b**) 0.25M $CaCl_2$

5.2.3 Release Kinetics Through Microcapsules

The pore size, the degradation rate, and ultimately the release kinetics of encapsulated biomedia are dependent on the different process parameters for alginate microcapsules [41]. The porous structure of the hydrogel can be tuned by controlling the degree of cross-linking by choosing appropriate concentrations of sodium alginate polymer and calcium chloride solutions. The gel matrix can be used to load drugs and their subsequent release at a rate dependent on the diffusion coefficient of the encapsulated biomolecule. Figure 5.11 shows a release kinetics profile of a fluorescent dye – Rhodamine 6G (R6G) that is encapsulated in a 20 μm diameter

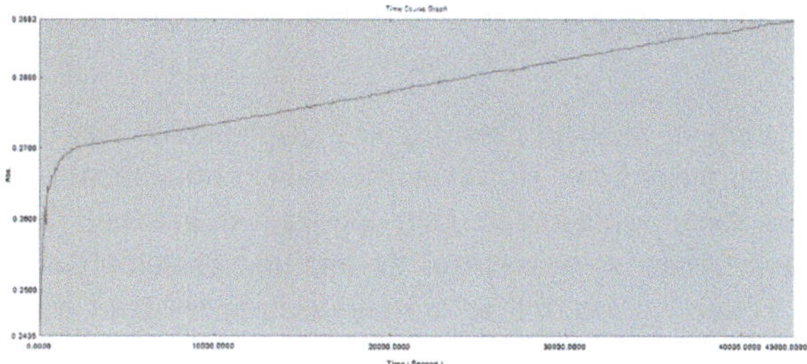

Fig. 5.11 Release profile of Rhodamine 6G dye from 20 μm dia biopolymer microcapsules (12-h profile, initial absorbance (0.2435), peak absorbance (0.2882) [107]

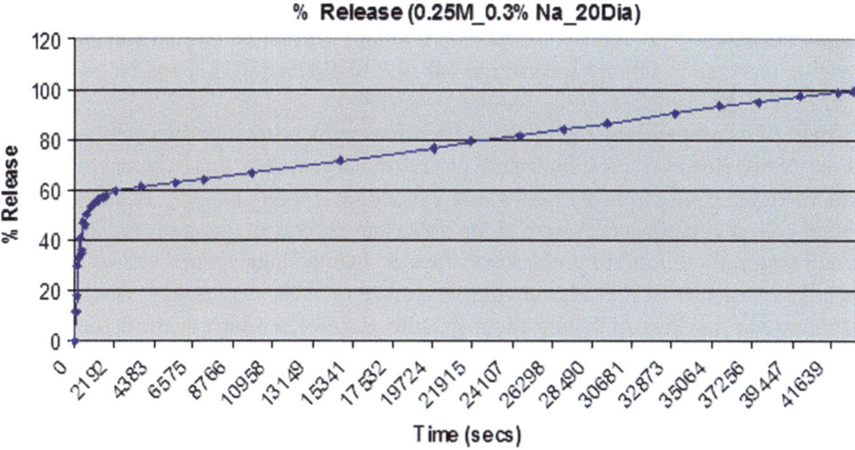

Fig. 5.12 Normalized profile of release kinetics of Rhodamine 6G dye from 20 μm dia biopolymer microcapsules (60% release of R6G during initial burst release (Courtesy: Perkins J., NC A&T SU)

hydrogel. The microcapsules were fabricated using the drop-on-demand inkjet printing of 0.3% (w/v) sodium alginate solution impregnated with R6G into 0.25M calcium chloride solution. Figure 5.12 depicts the percentage release profile of rhodamine 6G dye for a 12-h period. The initial burst release occurs over the first 1 h with approximately 60% release of the encapsulated R6G followed by a steady state release over the next 11 h. Different drug release profiles can be obtained by modifying the microcapsules diameter, biopolymer, and calcium chloride solution concentrations.

As the needs of the drug delivery systems grow more advanced, there will be a need to precisely control the release of drugs. Inkjetting may provide a useful method for producing uniform microbeads for drug encapsulation. The ability to print biopolymers such as sodium alginate and crosslink them into hydrogels readily demonstrates the feasibility of the inkjetting process for drug delivery systems.

5.2.4 Tissue Engineering

With advancements in medical care, organ transplantation has been a significant method for saving many lives. However, the reality is that there is a large shortage of donor organs for transplantation. One approach to alleviate this problem is by creating synthetic tissues or organs using the intelligent assembly of biological materials [42–44]. Tissue engineering is "a multidisciplinary field which involves the application of principles of engineering and life sciences toward the development of biological substitutes that restore, maintain or improve tissue function" [45, 46].

The basic paradigm of tissue engineering involves obtaining a small piece of donor tissue, isolating and expanding the cells, then attaching them to a suitable biomaterial or tissue scaffold followed by implantation. Over time the engineered tissue integrates into the body and completes the regeneration process. This paradigm has already proven to be viable in humans with tissue engineered constructs beging reported in skin, [47] cartilage, [48, 49] bladders, [50, 51] and blood vessels. [52–54]

The above mentioned reports have been for engineering two-dimensional or hollow organs. However, most biological tissues and organs have three-dimensional (3D) microscopic configuration of cells and extracellular matrices [55]. Besides assembling cells and biomaterials, one of the important aspects of tissue engineering is the combinatorial orchestrating of growth factors, extracellular matrix and cells within scaffold structures [56]. The unguided growth of cells results in a random two-dimensional stacking of cells without specific anatomical shape of the tissue [57].

To create three dimensional structures, several scaffold fabrication techniques exist that include solvent-casting particulate-leaching [58], gas foaming [59], fiber meshes/fiber bonding [60], phase separation [61], melt molding [62], and emulsion freeze drying [63] have been used. However, the degradation of synthetic polymers releases acidic by-products in conventional processes may not be ideal for tissue growth [64] nor can living cells usually be present during the construction of the biomaterial. In addition, many of the traditional scaffold fabrication methods are incapable of precisely controlling pore size, density, and internal conduits within the scaffold [65, 66]. Moreover, these processes have limited potential to control the distribution of impregnated nutrients, oxygen and growth factors to promote directed migration and cell differentiation within tissue engineering scaffolds [67–69]. Such disadvantages of current fabrication methods for tissue scaffolds may be overcome using the direct-write methods.

5.2.5 Direct-Write Methods for Tissue Engineering

For the successful production of tissue engineering scaffolds there are several factors that need to be considered [70]. The scaffold should posses interconnecting pores of varying sizes depending on the local structure of the tissue construct for appropriate vascularization with the host tissue. Secondly, the scaffold material should degrade within the body to complement the in-growth of new tissue constructs. It should have adequate mechanical strength and surface functionalizations to differentiate cell types for proliferation. Most importantly, the scaffold structure should be able to be fabricated in different shapes and sizes to adapt to the local tissue or organ configuration. Some of the direct-write based tissue fabrication technologies are discussed below.

5.2.5.1 Laser-Guided Direct-Writing (LGDW)

Laser-guided direct-writing (LGDW) is a non-contact patterning technique to directly write multiple cell lines and biological media using laser beams on arbitrary surfaces for tissue regeneration [71]. This technique has been demonstrated to pattern different cells in three-dimensional architecture with micrometer scale accuracy [72–74]. LGDW can embed oxygen enriched vascular structures within implanted tissue scaffolds. LGDW uses a weakly focused laser beam to generate functional optical forces with two components namely radial (orthogonal to beam axis) and axial in the direction of the beam propagation. Based on the optical forces particles can be drawn to the center of the beam and guided over tens or micrometers up to few millimeters onto substrates. The major components of the laser-guided direct-write system consist of laser source, the optics used to focus the beam for guidance, a cell deposition chamber where cell patterning occurs, a multiaxis stage for positioning the cell deposition chamber relative to the laser beam, a cell feeding mechanism, CCD cameras and imaging optics, and a computer controller [75]. The use of laser tweezer mechanisms for trapping individual cells and microparticles is an effective tool for investigating individual cell migration phenomena including cell–cell interactions and proliferation of cells within developing tissue [76, 77]. By controlling the motion of the underlying substrate relative to the laser beam one or more particles can be directly written onto the surface. The particles are driven to the substrate by optical forces which are based on the refractive index of the particles [78]. Different material particles including metal droplets, polymer beads, bacteria, and animal cells have been direct-guided to substrate locations [79]. However, from a product development and tissue engineering standpoint it has limited applicability due to lower throughput and longer cell placement times [80].

5.2.5.2 Extrusion-Based Direct-Writing of Colloidal Gels

This is a solid freeform fabrication (SFF) process that involves extrusion of specialized formulations of gels from nozzle to form a continuous filament on the substrate [81, 82]. Thus, three-dimensional periodic structures and arrays can be fabricated using robotic deposition of colloidal gels. The use of colloidal gels is a promising approach for building tissue engineering scaffolds as the viscoelastic properties of the polymers can be tuned for depositing different patterns [83]. However, nozzle clogging issues may be encountered for smaller nozzle diameters and limiting particle sizes [84]. Three-dimensional scaffold structures may be fabricated from bioactive ceramic colloidal gels [85] and the intermediate porosity can be infiltrated by appropriate growth factors [86]. Conventional solid freeform fabrication techniques have been used to create scaffolds with controlled porosities, material composition, and morphologies [87]. However, the harsh processing conditions including high temperatures, toxic chemicals, and radiation exposures can be detrimental for biofunctional structures [88]. Prof. Smay's Group at the Oklahoma State University has developed an aqueous, low-temperature direct-write method to deposit biomedia within scaffold structures for timed drug release [89]. In this process colloidal ink with polymer latex particles are extruded from nozzle to generate filament structures on the substrate. By manipulating the substrate relative to the nozzle three-dimensional structures are formed by stacking individual 2D patterns layer by layer. This process allows for fabrication of scaffold structures at biologically conducive conditions with adequate mechanical strength. The aqueous suspensions can be infiltrated with relevant growth factors that promote cell adhesion and proliferation [90]. The Lewis Group at the University of Illinois, Urbana-Champaign [91–93] has developed novel formulations of inks including concentrated colloids [94], nanoparticle [95], fugitive organic [96] and polyelectrolyte inks [97] capable of direct-writing complex three-dimensional structures that can be applied toward tissue engineering.

5.2.5.3 Inkjet-Based Tissue Engineering

Inkjet printing provides a viable method to selective deposit the individual biomedia to target-specific locations. This method provides fewer limitations for transferring biological materials to substrates in extremely small volumes. Different types of proteins [98, 99], growth factors [100], and complete cell lines [101, 102] have been deposited using the inkjet technology. Multiple inkjet nozzles can be used to deposit different cell types in anatomically relevant positions to mimic realistic tissue structures. By manipulating the underlying substrates using computer-aided tools it is possible to print human and animal cells in desired patterns to regenerative new tissue or organ [103]. Complex three-dimensional structures which simultaneously print cells and biomaterials such as shown in Fig. 5.13.

Multipotent neural stem cells have been inkjet printed to differentiate into neural and non-neural linkages [104, 105] with the addition of extrinsic factors for replacement of cells lost due to disease or injury [106]. In addition, Fig. 5.14 demonstrates that the printed stem cells can also be differentiated into bone. Such results show that the inkjet printing process is capable of printing cells that retain their normal physiological properties.

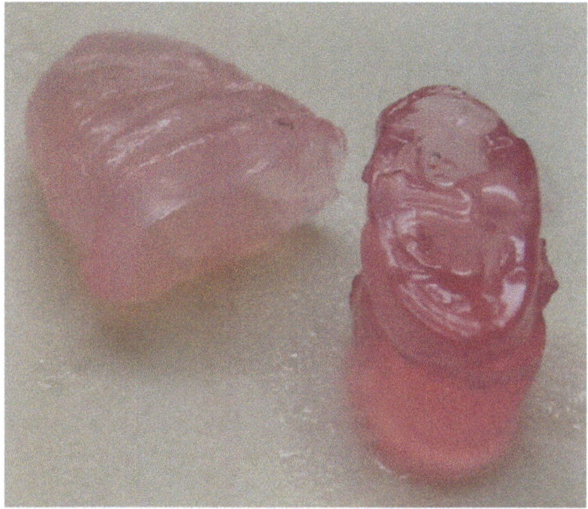

Fig. 5.13 A complex 3D structure prepared through thermal inkjetting which allows simultaneous incorporation of cells into the tissue scaffold

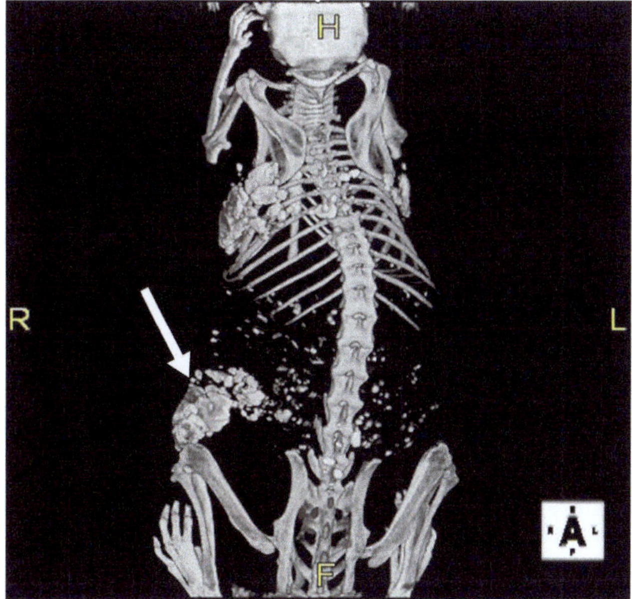

Fig. 5.14 CT image of an inkjet printed construct seeded with stem cells and osteogenic growth factors. The *white arrow* shows bone formation at the site of the implanted construct. (Courtesy of Xu T., WFIRM.)

References

1. McAllister DV et al (2000) Microfabricated microneedles for gene and drug delivery. Annu Rev Biomed Eng 2:289–313
2. Gao X et al (2004) In vivo cancer targeting and imaging with semiconductor quantum dots. Nat Biotechnol 22(8):969–976
3. Langer R, Tirrell DA (2004) Designing materials for biology and medicine. Nature 428(6982):487–492
4. Bredt JF, Sach E, Brancazio D, Cima M, Curodeau A, Fan T (1998) Three dimensional printing system. US Patent 5807437
5. Geng L, Feng W, Hutmacher DW, Wong YS, Loh HT, Fuh JYH (2005) Direct writing of chitosan scaffolds using a robotic system". Rapid Prototyping J 11(2):90–97
6. McGuigan AP, Sefton MV (2006) Vascularized organoid engineered by modular assembly enables blood perfusion. Proc Natl Acad Sci USA 103(31):11461–11466
7. Sengupta S et al (2005) Temporal targeting of tumour cells and neovasculature with a nanoscale delivery system. Nature 436(7050):568–572
8. Ilkhanizadeh S, Teixeira AI, Hermanson O (2007) Inkjet printing of macromolecules on hydrogels to steer neural stem cell differentiation. Biomaterials 29:3936–3943
9. Langer R (1998) Drug delivery and targeting. Nature 392:5–10
10. Jacobson JM, Hubert BN, Ridley B, Nivi B, Fuller S. Nanoparticle-based electrical, chemical, and mechanical structures and methods of making same. United States Patent: 6,294,401, September 25, 2001.
11. Szczech JB, Megaridis CM, Gamota DR, Zhang J (2002) Fine-line conductor manufacturing using drop-on-demand PZT printing technology. IEEE Transactions on Electronics Packaging Manufacturing 25(1):26–33
12. Garnier F, Hajlaoui R, Yassar A, Srivastava P (1994) All-polymer field-effect transistor realized by printing techniques. Science Magazine 265:1684–1686
13. Ridley BA, Nivi B, Jacobson JM (1999) All Organic Field Effect Transistors Fabricated by Printing. Science Magazine 286:746–748
14. Shah VG, Hayes DJ (2002) Fabrication of passive elements using ink-jet technology. IMAPS ATW on Passive Integration, MicroFab Technologies, Inc.
15. Lemmo AV, Rose DJ, Tisone TC (1998) Inkjet dispensing technology: applications in drug discovery. Curr Opin Biotechnol 9(6):615–617
16. Boland T, Xu T, Damon B, Cui X (2006) Application of inkjet printing to tissue engineering. Biotechnol J 1(9):910–917
17. Leslie M (2006) Sprayed-on growth factors guide stem cells. Science 314(5807):1865
18. Alper J (2004) Biology and the inkjets. Science 305(5692):1895
19. Lausted CG, Warren CB, Hood LE, Lasky SR (2006) Printing your own inkjet microarrays. Methods Enzymol 410:168–189
20. Le HP (1999) Progress and trends in ink-jet printing technology, Recent progresses in inkjet technologies II, The Society for Imaging Science and Technology. 1–14.
21. Rayleigh FRS (1878) On the instability of Jets. Proc Lond Math Soc 10(4):4–13
22. Yatsuzuka K, Saito K, Asano K, Aero-dynamical motion of charged droplets ejected from the 26 μm Nozzle. IEEE Industry Applications Society Annual Meeting, New Orleans, October 1997, pp. 1845–1850
23. Bogy DB, Talke FE (1984) Experimental and theoretical study of wave propagation phenomena in drop-on-demand ink jet devices Volume 28. IBM J Res Dev 28:314–321
24. Roth EA, XuT DM, Gregory C, Hickman JJ, Boland T (2004) Inkjet printing for high-throughput cell patterning. Biomaterials 25(17):3707–3715
25. Roda A, Guardigli M, Russo C, Pasini P, Baraldini M (2000) Protein microdeposition using a conventional ink-jet printer. Biotechniques 28(3):492–496
26. Okamoto T, Suzuki T, Yamamoto N (2000) Microarray fabrication with covalent attachment of DNA using bubble jet technology. Nat Biotechnol 18(4):438–441

27. Nakamura M, Kobayashi A, Takagi F, Watanabe A, Hiruma Y, Ohuchi K, Iwasaki Y, Horie M, Morita I, Takatani S (2005) Biocompatible inkjet printing technique for designed seeding of individual living cells. Tissue Eng 11(11/12):1658–1666
28. de Gans B-J, Duineveld PC, Schubert US (2004) Inkjet printing of polymers: state of the art and future developments. Adv Mater 16(3):203–213
29. Khalil SED (2005) Deposition and Structural Formation of 3D Alginate Tissue Scaffolds. Drexel University, Philadelphia, Ph.D. Dissertation
30. Richards Grayson AC et al (2003) Multi-pulse drug delivery from a resorbable polymeric microchip device. Nat Mater 2(11):767–772
31. Wichterle O, Lim D (1960) Hydrophilic gels for biological use. Nature 185:117–118
32. Dong Y (2007) In vitro and in vivo evaluation of methoxy polyethylene glycol–polylactide (MPEG–PLA) nanoparticles for small-molecule drug chemotherapy. Biomaterials 28:4154–4160
33. Murtas S, Capuani G, Dentini M, Manetti C, Masci G, Massimi M et al (2005) Alginate beads as immobilization matrix for hepatocytes perfused in a bioreactor: a physico-chemical characterization. J Biomater Sci Polym Ed 16(7):829–846
34. Sutton C (2005) The obstetrician and gynaecologist 7:168e76
35. Giunchedi P, Gavini E, Moretti MDL, Pirisino G (2000) Evaluation of alginate compressed matrices as prolonged drug delivery systems. AAPS Pharm Sci Tech 1:31–36
36. Clark AH, Ross-Murphy SB (1987) Adv Polym Sci 83:57
37. Gombotz WR, Wee SF (1998) Protein release from alginate matrices. Adv Drug Deliv Rev 31:267–285
38. Ciofani G, Raffa V, Menciassi M, Dario P (2008) Alginate and chitosan particles as drug delivery system for cell therapy. Biomed Microdevices 10:131–140
39. Douglas KL, Tabrizian M (2005) Effect of experimental parameters on the formation of alginate–chitosan nanoparticles and evaluation of their potential application as DNA carrier. J Biomater Sci Polym Ed 16(1):43–56
40. Moore A (2006) Understanding microdrop fluid generation of a biopolymer system for biomedical applications. Masters Thesis, Department of Industrial and Systems Engineering, NC A&T SU, Oct 2006
41. Wang MS, Childs RF, Chang PL (2005) A novel method to enhance the stability of alginate-poly-l-lysine-alginate microcapsules. J Biomater Sci Polym Ed 16(1):89–111
42. Kaihara S, Vacanti JP (1999) Tissue engineering: toward new solutions for transplantation and reconstructive surgery. Arch Surg 134:1184
43. Griffith LG, Naughton G (2002) Tissue engineering: current challenges and expanding opportunities. Science 295:1009
44. Fuchs JR, Nasseri BA, Vacanti JP (2001) Tissue engineering: A 21st century solution to surgical reconstruction. Ann Thorac Surg 72:577
45. Shalak R, Fox CF (1988) Preface. In: Tissue Engineering. Alan R. Liss (eds), New York. pp 26–29
46. Langer R, Vacanti JP (1993) Tissue Engineering. Science. American Assocation for the advancement of Science (AAAS)260(5110):920–926
47. Gentzkow GD, Iwasaki SD, Hershon KS, Mengel M, Prendergast JJ, Ricotta JJ et al (1996) Use of dermagraft, a cultured human dermis, to treat diabetic root ulcers. Diabetes Care 19:350–354
48. Brittberg M, Lindahl A, Nilsson A, Ohlsson C, Isaksson O, Peterson L (1994) Treatment of deep cartilage defects in the knee with autologous chondrocyte transplantation. N Engl J Med 331:889–895
49. Caldamone AA, Diamond DA (2001) Long-term results of the endoscopic correction of vesicoureteral reflux in children using autologous chondrocytes. J Urol 165:2224–2227
50. Atala A, Bauer SB, Soker S, Yoo JJ, Retik AB (2006) Tissue-engineered autologous bladders for patients needing cystoplasty. Lancet 367:1241–1246
51. Oberpenning F, Meng J, Yoo JJ, Atala A (1999) De novo reconstitution of a functional mammalian urinary bladder by tissue engineering. Nat Biotechnol 17:149–155
52. Kaushal S, Amiel GE, Guleserian KJ, Shapira OM, Perry T, Sutherland FW et al (2001) Functional small-diameter neovessels created using endothelial progenitor cells expanded ex vivo. Nat Med 7:1035–1040

53. Shin'oka T, Matsumura G, Hibino N, Naito Y, Watanabe M, Konuma T et al (2005) Midterm clinical result of tissue-engineered vascular autografts seeded with autologous bone marrow cells. J Thorac Cardiovasc Surg 129:1330–1338
54. Matsumura G, Hibino N, Ikada Y, Kurosawa H, Shin'oka T (2003) Successful application of tissue engineered vascular autografts: clinical experience. Biomaterials 24:2303–2308
55. Gartner L, Hiatt J (1997) Color textbook of histology, 2nd edn. W. B. Saunders, Philadelphia, PA, p 11
56. Sun W, Darling A, Starly B, Nam J (2003) Computer aided tissue engineering Part I: Overview, scope and challenges. Biotechnol Appl Biochem 39 9pt 1: 29–47
57. Sachlos E, Czernuszka JT (2003) Making tissue engineering scaffolds work. Review on the application of solid freeform fabrication technology to the production of tissue engineering scaffolds. Eur Cell Mater 5:29–40
58. Mikos AG, Thorsen AJ, Czerwonka LA, Bao Y, Langer R (1994) Preparation and characterisation of poly(L-lactic acid) foams. Polymer 35:1068–1077
59. Mooney DJ, Baldwin DF, Suh NP, Vacanti JP, Langer R (1996) Novel approach to fabricate porous sponges of poly(D, L-lactic co-glycolic acid) without the use of organic solvents. Biomaterials 17:1417–1422
60. Cima LG, Vacanti JP, Vacanti C, Inger D, Mooney D, Langer R (1991) Tissue engineering by cell transplantation using degradable polymer substrates. J Biomech Eng-T ASME 113:143–151
61. Lo H, Ponticiello MS, Leong KW (1995) Fabrication of controlled release biodegradable foams by phase separation. Tissue Eng 1:15–28
62. Thompson RC, Yaszemski MJ, Powers JM, Mikos AG (1995) Fabrication of biodegradable polymer scaffolds to engineering trabecular bone. J Biomater Sci-Polym E 7:23–38
63. Whang K, Thomas CK, Nuber G, Healy KE (1995) A novel method to fabricate bioabsorbable scaffolds. Polymer 36:837
64. Reed AM, Gilding DK (1981) Biodegradable polymer for use in surgery –poly(glycolic)/ poly(lactic acid) homo and copolymers: 2. In vitro degradation. Polymer 22:342–346
65. Ishaug-Riley SL, Crane GM, Gurlek A, Miller MJ, Yasko AW, Yaszemski MJ, Mikos AG (1997) Ectopic bone formation by marrow stromal osteoblast transplantation using poly(DL-lactic-co-glycolic acid) foams implanted into the rat mesentery. J Biomed Mater Res 36:1–8
66. Freed LE, Vunjak-Novakovic G (1998) Culture of organized cell communities. Adv Drug Deliver Rev 33:15–30
67. Martin I, Padera RF, Vunjak-Novakovic G, Freed LE (1998) In vitro differentiation of chick embryo bone marrow stromal cells into cartilaginous and bone-like tissues. J Orthopaed Res 16:181–189
68. Vander AJ, Sherman JH, Luciano DS (1985) Human Physiology. McGraw-Hill, New York, pp 341–366
69. Eaglstein WH, Falanga V (1997) Tissue engineering and the development of Apligraf®, a human skin equivalent. Clin Ther 19:894–905
70. Hutmacher DW (2001) Scaffold design and fabrication technologies for engineering tissues-state of the art and future perspectives. J Biomat Sci-Polym E 12:107–124
71. Yaakov Nahmias, Schwartz RE, Verfaillie CM, Odde DJ (2005) Laser-guided direct writing for three-dimensional tissue engineering. Biotechnology and Bioengineering. 92, No. 2.
72. Chrisey DB (2000) The power of direct writing. Science 289(5481):879–881
73. Nahmias YK, Gao BZ, Odde DJ (2004) Dimensionless parameters for the design of optical traps and laser guidance systems. Appl Opt 43(20):3999–4006
74. Odde DJ, Renn MJ (1999) Laser-guided direct writing for applications in biotechnology. Trends Biotechnol 17:385–389
75. Narasimhan SV, Goodwin RL, Borg RL, Dawson TK et al (2004) Multiple beam laser cell micropatterning system. Proc. SPIE 5514:437–446
76. Wright WH, Sonek GJ, Tadir Y, Berns MW (1990) Laser trapping in cell biology. IEEE J Quantum Elect 26:2148–2157
77. Neuman KC, Block SM (2004) Optical trapping. Rev Sci Instrum 75:2787–2809
78. Odde DJ, Renn MJ (2000) Laser-guided direct writing of living cells. Biotechnol Bioeng 67:312–318

79. Ashkin A (2000) History of optical trapping and manipulation of small neutral particle, atoms, and molecules. IEEE J Sel Top Quant 6:841–856
80. Pirlo RK, Delphine MD, Knapp DR, Gao BZ (2006) Cell deposition system based on laser guidance. Biotechnol J 1:1007–1013
81. Smay JE et al (2002) Langmuir 18(14):5429
82. Smay JE et al (2002) Adv Mater 14(18):1279
83. Lewis JA, Gratson GM (2004) Direct writing in three dimensions. Materials Today 7:32–39
84. Li Q, Lewis JA (2003) Adv Mater 15(19):1639
85. Chu TMG et al (2001) J Mater Sci: Mater Med 12(6):471
86. Dhariwala B, Hunt E, Boland T (2004) Tissue Eng 10:1316
87. Leong KF, Cheah CM, Chua CK (2003) Biomaterials 24:2363
88. Landers R, Pfister A, Huebner U, John H, Schmelzeisen R, Muelhaupt R (2002) J Mater Sci 37:3107
89. Xie B, Parkhill RL, Warren WL, Smay JE (2006) Direct writing of three-dimensional polymer scaffolds using colloidal gels. Adv Funct Mater 16:1685–1693
90. Janes KA, Alonso MJ (2003) J Appl Polym Sci 88:2769
91. Chan AT, Lewis JA (2005) Electrostatically tuned interactions in silica microsphere-polystyrene nanoparticle mixtures. Langmuir 21:8576–8579
92. Rhodes SK, Lewis JA (2006) Phase behavior, 3-D structure, and rheology of colloidal microsphere-nanoparticle suspensions. J Am Ceram Soc 89:1840–1846
93. A Mohraz ER, Weeks JA Lewis (2008) Structure and dynamics of biphasics colloidal mixtures. Physical Review E 77, 060403(R)
94. George MC, Mohraz A, Piech M, Bell NS, Lewis JA, Braun PV (2008) Direct laser patterning of photoresponsive colloids for microscale patterning of 3D porous structures. Adv Mater 20:1–5
95. Lee W, Chan A, Lewis JA, Braun PV (2004) Nanoparticle-mediated epitaxial assembly of colloidal microspheres on patterned substrates. Langmuir 20:5262–5270
96. Therriault D, White S, Lewis JA (2007) Rheological behavior of fugitive organic inks for direct-write assembly. Appl Rheol 17; 10112-1–10112-8
97. Gratson G, Lewis JA (2005) Polyelectrolyte inks for direct-write assembly of 3-D microperiodic scaffolds. Langmuir 21:457–464
98. Roth EAXuT, Das M, Gregory C, Hickman JJ, Boland T (2004) Inkjet printing for high-throughput cell patterning. Biomaterials 25(17):3707–3715
99. Sanjana NE, Fuller SB (2004) A fast flexible ink-ket printing method for patterning dissociated neurons in culture. J Neurosci Meth 136(2):151–163
100. Watanable K, Miyazaki T, Matsuda R (2003) Growth factor array fabrication using a color ink jet printer. Zool Sci 20(4):429–434
101. Xu T, Petridou S, Lee EH, Roth EA, Vyavahare NR, Hickman JJ et al (2004) Construction of high-density bacterial colony arrays and patterns by the ink-jet method. Biotechnol Bioeng 85(1):29–33
102. Xu T, Jin J, Gregory C, Hickman JJ, Boland T (2005) Inkjet printing of viable mammalian cells. Biomaterials 26:93–99
103. Mironov V, Boland T, Trusk T, Forgacs G, Markwald RR (2003) Organ printing: computer-aided jet-based 3D tissue engineering. Trends Biotechnol 21(4):157–161
104. Johe KK, Hazel TG, Muller T, Dugich-Djordjevic MM, McKay RD (1996) Single factors direct the differentiation of stem cells from the fetal and adult central nervous system. Genes Dev 10(24):3129–3140
105. Hermanson O, Jepsen K, Rosenfeld MG (2002) N-CoR controls differentiation of neural stem cells into astrocytes. Nature 419(6910):934–939
106. Teixeira AI, Duckworth JK, Hermanson O (2007) Getting the right stuff: controlling neural stem cell state and fate in vivo and in vitro with biomaterials. Cell Res 17(1):56–61
107. Perkins J (2008) Characterization of absorbance and time release kinetics of calcium alginate microcapsules using drop-on-demand inkjet printing technology" Masters Thesis, Department of Industrial & Systems Engineering, NC A&T SU, Dec 2008

Chapter 6
Precision Extruding Deposition for Freeform Fabrication of PCL and PCL-HA Tissue Scaffolds

L. Shor, E.D. Yildirim, S. Güçeri, and W. Sun

Abstract Computer-aided tissue engineering approach was used to develop a novel Precision Extrusion Deposition (PED) process to directly fabricate Polycaprolactone (PCL) and composite PCL/Hydroxyapatite (PCL-HA) tissue scaffolds. The process optimization was carried out to fabricate both PCL and PCL-HA (25% concentration by weight of HA) with a controlled pore size and internal pore structure of the 0°/90° pattern. Two groups of scaffolds having 60 and 70% porosity and with pore sizes of 450 and 750 microns, respectively, were evaluated for their morphology and compressive properties using Scanning Electron Microscopy (SEM) and mechanical testing. The surface modification with plasma was conducted on PCL scaffold to increase the cellular attachment and proliferation. Our results suggested that inclusion of HA significantly increased the compressive modulus from 59 to 84 MPa for 60% porous scaffolds and from 30 to 76 MPa for 70% porous scaffolds. In vitro cell–scaffolds interaction study was carried out using primary fetal bovine osteoblasts to assess the feasibility of scaffolds for bone tissue engineering application. In addition, the results in surface hydrophilicity and roughness show that plasma surface modification can increase the hydrophilicity while introducing the nano-scale surface roughness on PCL surface. The cell proliferation and differentiation were calculated by Alamar Blue assay and by determining alkaline phosphatase activity. The osteoblasts were able to migrate and proliferate over the cultured time for both PCL as well as PCL-HA scaffolds. Our study demonstrated the viability of the PED process to the fabricate PCL and PCL-HA composite scaffolds having necessary mechanical property, structural integrity, controlled pore size and pore interconnectivity desired for bone tissue engineering.

Keywords Bone tissue engineering • Polycaprolactone • Hydroxyapatite • Composite • Free form fabrication • Precision extrusion deposition • Plasma surface modification

---

L. Shor, E.D. Yildirim, S. Güçeri, and W. Sun (✉)
Laboratory for Computer-Aided Tissue Engineering, Department of Mechanical Engineering and Mechanics, Drexel University, Philadelphia, PA, 19104, USA
e-mail: sunwei@drexel.edu

R. Narayan et al. (eds.), *Printed Biomaterials*, Biological and Medical Physics, Biomedical Engineering, DOI 10.1007/978-1-4419-1395-1_6,

6.1 Introduction

The scaffolds designed for tissue engineering applications should be three-dimensional, highly porous, and interconnected to support cell attachment as well as proliferation. They should have sufficient structural integrity matching the mechanical properties of native tissue. They should provide suitable pore size distribution for transportation of nutrients and wastes. The scaffolds should offer ideal and critical microenvironment so that they can function as an artificial extracellular matrix (ECM) onto which cells attach, grow, and form new tissues [1–3]. Most available scaffold fabrication methods, such as solvent casting, fiber bonding, phase separation, gas-induced foaming, and salt leaching, are either limited to producing scaffolds with simple geometry, or depend on in-direct casting method for scaffold fabrication [4, 5], therefore they are impractical for the manufacturing of scaffolds with complex structural architectures. These traditional scaffold fabrication methods result in structures of random internal architecture and have great variation from part to part. Various Solid Freeform Fabrication (SFF) techniques including 3D Printing, Selective Laser Sintering, Multiphase Jet Solidification, and Fused Deposition Modeling (FDM) have been used successfully to manufacture advanced tissue scaffolds with specific designed properties [6–11]. The scaffolds manufactured using SFF methods have 100% interconnectivity and the porosity of these scaffolds can easily be controlled by optimizing the processing parameters. The SFF technique offers a unique opportunity to study the influence of the microarchitecture on cell proliferation and ECM generation. The computer-aided tissue engineering method can be used to create scaffolds that both incorporate patient-specific information as well as an explicitly designed microenvironment. Tissue geometry can be extracted from patient's Computed Tomography (CT) or Magnetic Resonance Imaging (MRI) data and reconstructed as a 3D model. Furthermore, as with most computer-aided design, detail analysis of the mechanical and transport properties can aid in the understanding of tissue growth in a scaffold-guided environment.

Among different SFF methods, FDM has recently attracted more interests due to its ability to form 3D structures by layer-by-layer deposition. The system utilizes a filament of thermoplastic material that is fed into a liquefying chamber by two rollers. These rollers provide the necessary pressure to extrude the molten material out through a nozzle tip. However, the time consuming precursor step of filament fabrication acts as a main obstacle for FDM [12]. Furthermore, with brittle materials frequent filament buckling failures during the extrusion of material cause interruption of the process and necessitate numerous operator interventions [13, 14] thereby limiting the available materials. Consequently, this problem prevents an automatic and continuous process diminishing the main advantage of a filament-based system. To alleviate this problem we developed a new system called Precision Extrusion Deposition (PED) consisting of a mini-extruder mounted on a high-precision positioning system (Fig. 6.1). PED can be used with bulk material in granulated form, which avoids most of the material preparation steps in a filament-based system.

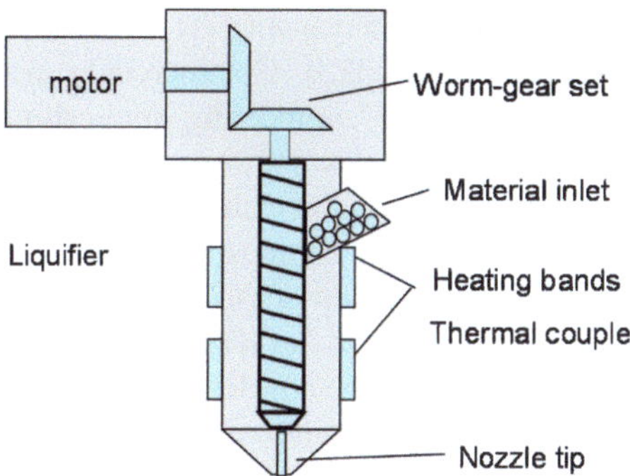

Fig. 6.1 Schematic of mini-extruder in precision extrusion deposition system

This configuration opens up the opportunity for the use of a wider range of materials, making the PED a viable alternative manufacturing process for composite scaffold materials.

Because of the advances in the scaffolds fabrication techniques bone tissue engineering is increasingly becoming a method of choice for the development of viable substitutes for skeletal reconstruction. Material used for fabricating scaffolds for bone tissue engineering application should have the mechanical integrity sufficient enough for bone cells to attach, proliferate, and differentiate in a manner similar to native ECM. Polycaprolactone (PCL) has been used by many for such an application because of its biodegradable and biocompatible properties [2, 9, 10]. PCL is a semi-crystalline aliphatic polymer that has a slower degradation rate than most biopolymers in its homo-polymeric form. It has a low glass transition temperature at −60°C, a melting temperature at about 58–60°C, and a high thermal stability. It has a high decomposition temperature of 350°C. The mechanical properties of bulk PCL (Mw=44,000) with a tensile strength of 16 MPa, tensile modulus of 400 MPa, flexural modulus of 500 MPa, elongation at an yield of 7%, and elongation at a break of 80% have been reported [15–20]. Nevertheless, the aforementioned properties of PCL make it an ideal biomaterial for bone tissue scaffold and PED scaffold manufacturing system, its bioinertness originates some problems with the cell interactions [21]. Surface modification and incorporation of bioactive materials to PCL can be an approach to promote the quality of cell–biomaterial interactions.

In literature, the plasma surface modification of biopolymer is widely used to increase the cellular functions on scaffold. Depending on the working gas (oxygen, nitrogen, air, etc) used in plasma formation, the physiochemical properties of biomaterial can be altered without changing its bulk properties. In addition, plasma surface modification introduces nano-scale surface roughness to the surface while changing the surface chemical composition of the material [22]. Besides surface modification of PCL, another way to increase the cellular functions on PCL scaffold is to incorporate bioactive material, hydroxyapatite (HA), into PCL. Chemically, bone is made up of 58% calcium phosphate, 7% calcium carbonate, 1–2% calcium fluoride, 1–2% magnesium phosphate, and 1% sodium chloride. These minerals together form a crystal called hydroxyapatite (HA). Remaining part of bone is made up of water, cells, and ECM [23]. Commercially available HA has been widely investigated for bone tissue engineering applications [9–11]. It is both mechanically strong and osteoconductive. However, because of its brittle characteristic and material properties, it is often difficult to process.

We fabricated PCL and PCL-HA composite scaffolds, having 25% HA by weight using PED system. Then, a group of PCL scaffold was exposed to oxygen-based plasma for 3 min to modify their surfaces. The pore size and porosity were optimized by varying the diameter of the nozzle tip. Scanning Electron Microscopy (SEM) was used to characterize the morphologies and microstructures of the PED-fabricated scaffolds. Instron 5800R was used to calculate mechanical property of scaffolds. The cell–scaffold interaction was studied using primary fetal bovine osteoblasts.

6.2 Scaffold Fabrication

PCL (Sigma Aldrich Inc., Milwaukee, WI) in the form of pellets was used as the scaffolding material. Hydroxyapatite (Clarkson Chromatography Products Inc., South Williamsport, PA) in a form of a powder, with particles ranging in size from 10 to 25 microns, was melt blended with PCL, with 25% HA by weight for the fabrication of composite scaffold. PED system developed at Drexel University [24] was used for manufacturing scaffolds. The mini-extruder system (Fig. 6.1) delivers the PCL or PCL-HA in a fused form through the deposition nozzle. The material is fused by a liquefier temperature provided by two heating bands and respective thermal couples. PCL or PCL-HA is then extruded due to pressure created by turning precision screw. Two sets of cylindrical scaffolds, measuring 20 mm in diameter, were fabricated with 450 micron struts (width of the extruded material), and porosities of 60 and 70%, respectively. The liquefier temperature was set to 90°C, and a 0.245 mm exit diameter nozzle was used. Each layer was filled with the designed scaffold pattern of a 0°/90° orientation to generate the porous structure.

6.2.1 Porosity of Scaffolds

Figure 6.2 shows the model of the scaffold with 0°/90° layout pattern. By assuming that the pore spacing is consistent throughout the entire height of the scaffold, the porosity was calculated from imaging the top surface strut pattern [15]. After the fabrication the top surface of the scaffold, the pore size and extrudate width were imaged and measured using a microscope (Leica DM IL). In Fig. 6.2, $2L$ is the length between the struts which is half of the gap length (G), α is the orientation angle and D is the diameter of the strut. According to Fig. 6.2, the volume of the scaffold, V, can be calculated as:

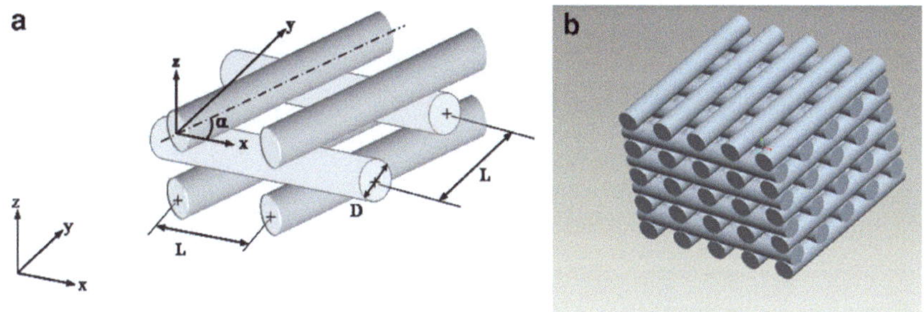

Fig. 6.2 Schematic of porosity calculation (**a**) and strand layout (**b**)

V = 2L * D

The volume of the strut can be calculated by:

$$V_{\text{rode}} = \frac{1}{4}\frac{\pi D^2}{4}L + \frac{1}{4}\frac{\pi D^2}{4}\frac{L}{\sin\alpha} = \frac{\pi D^2}{16}\left(L + \frac{L}{\sin\alpha}\right)$$

Note that the volume is calculated by taking a quarter of each strut in the scaffold since the repetitive unit consists of quarter struts. Therefore, the volume fraction of the strut in the scaffold can be determined as:

$$v_f = \frac{V_{\text{rode}}}{V} = \frac{\pi D^2}{32L\ D}\left(L + \frac{L}{\sin\alpha}\right) = \frac{\pi D}{16}\left(\frac{1}{L} + \frac{1}{L\sin\alpha}\right)$$

The porosity (φ) of the scaffolds can then be calculated by:

$$\phi = 1 - v_f = 1 - \frac{\pi D}{16L}\left(1 + \frac{1}{\sin\alpha}\right)$$

6.2.2 *Morphology Study by SEM*

FEI/Phillips XL-30 Field Emission Environmental Scanning Electron Microscope was used to evaluate the morphologies of PCL and PCL-HA scaffolds. The SEM images were taken by using beam intensity at 20 kV and the gaseous secondary electron detectors at 1.3 Torr (Fig. 6.3).

The gaps, struts, and internal pore connectivity, as observed under SEM, demonstrate use of the PED process to fabricate PCL scaffolds at the micro-scale level. The SEM images clearly demonstrate that the PED-fabricated micro-architecture of the scaffolds via a 0°/90° layered pattern achieves the desired pore size of 350 μm.

The backscatter SEM images, showing HA particle in white, of melt blended PCL-HA (Fig. 6.4a) and as fabricated scaffold (Fig. 6.4b) were taken to examine

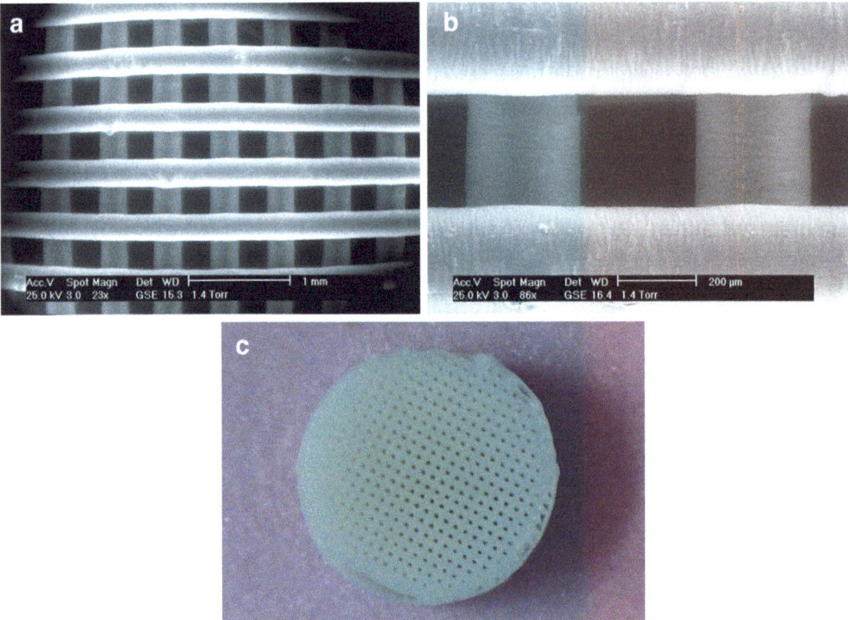

Fig. 6.3 Scanning electron microscope images of scaffold showing 0°/90° layout pattern. Figure (**a**) and (**b**) represent lower and higher magnifications, respectively. Figure (**c**) shows an unmagnified image of the scaffold

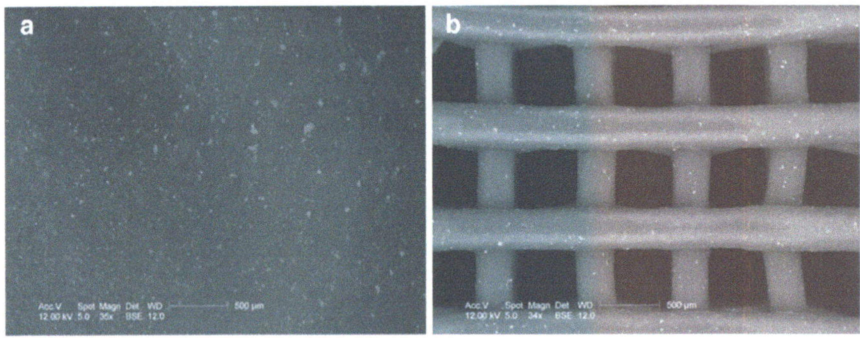

Fig. 6.4 Backscatter scanning electron microscopy images of melt blended PCL-HA (**a**) and as fabricated scaffold (**b**)

the HA distribution in PCL. It shows well-dispersed HA particles within PCL with no visible areas of agglomeration. The required architecture for tissue engineering scaffolds could be achieved at the micron-scale level. The uniformity of the pores and the depositing roads shown demonstrate the applicability of using the PED process to fabricate composite scaffolds at micro-scale level.

6.3 Mechanical Properties of Scaffolds

Compression tests were conducted on both PCL and PCL-HA (25% concentration by weight) scaffolds. The Instron 5800R machine was used for evaluating compressive properties. The samples were 20 mm in diameter and 20 mm in height. Scaffolds of 60 and 70% porosity with pore sizes of 450 and 750 microns, respectively, were tested to determine the effect of porosity on mechanical properties. The tests were conducted with a cross-head displacement speed of 2 mm/min. Stress–Strain data was computed from Load–Displacement measurements and the compressive modulus was determined from the elastic region of the curve.

The stress–strain curves derived from the testing data are plotted in Fig. 6.5. The calculated compressive modulus from the stress–strain data are listed in Table 6.1. The results show that the less porous scaffolds had overall better properties. Inclusion of HA increased the compressive modulus from 59 to 84 MPa for 60% porous scaffolds and from 30 to 76 MPa for 70% porous scaffolds.

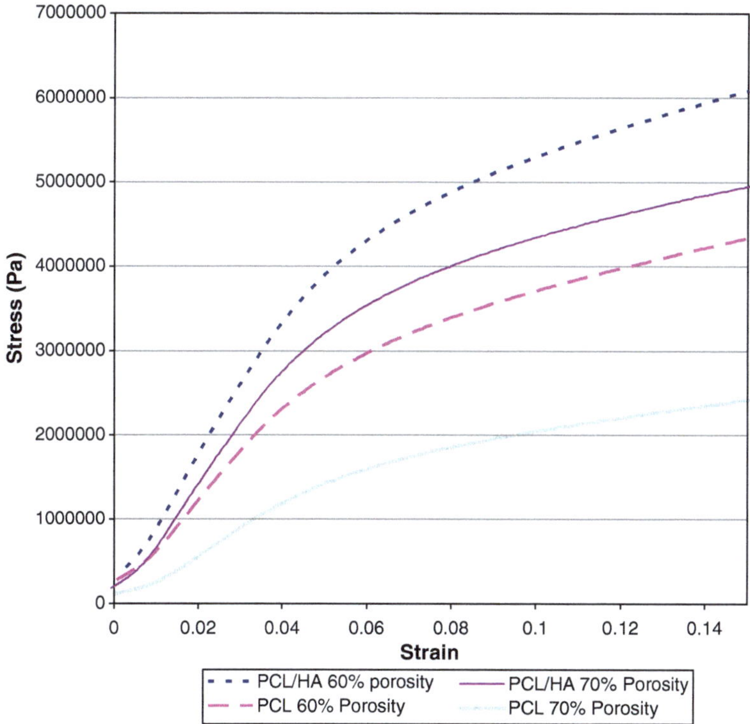

Fig. 6.5 Stress–strain curve for PCL and composite PCL/HA scaffold

Table 6.1 Calculated compressive modulus from stress–strain data

Material	Porosity (%)	Compressive modulus (MPa)
PCL-HA	60	84
PCL	60	59
PCL-HA	70	76
PCL	70	30

6.4 Surface Modification of PCL Scaffolds

The surfaces of PCL scaffolds were modified with a plasma reactor (PDC 32G, Harrick Scientific Inc., New York). The system included a radiofrequency generator capable of 0–18 W at frequency range of 8–12 MHz, a vacuum pump, a helical internal electrode around the reactor, and instrumentation for pressures. 3D PCL scaffolds were placed inside the chamber and exposed to the plasma for 3 min at 10 psi with a pure oxygen gas flow rate of 1 standard liter/min (slpm) and power of 18 W at room temperature. After the plasma treatment, the surface characterization and biological characterization were conducted to analyze the effect of oxygen-based plasma modification on surface physiochemical properties and cell–scaffold interaction. The surface characterization was done by measuring the surface hydrophilicity and energy using contact angle measurement of three different probe liquids and by quantifying the surface roughness using atomic force microscopy (AFM).

6.4.1 Surface Hydrophilicity and Energy

The contact angle measurements were used to evaluate the effect of oxygen-based plasma treatment on PCL surface in terms of degree of hydrophilicty and solid surface energy. The contact angle (θ) of probe liquids on 3-min modified and unmodified PCL surfaces were measured by sessile drop technique. As probe liquid, diiodomethane (Fisher, PA), glycerol (Fisher, PA), and ultra pure water (Agilent, Germany) were used as a fact that pairs of polar and apolar liquids have to be used to obtain reliable values of solid surface energy. Drop of probe liquid ($2\,\mu$L) was placed onto plasma-modified and control PCL sample surface. When the liquid has settled (become sessile), contact angle measurements were taken at least four times to obtain a grand average. All contact angle measurements were done at room temperature.

The results from contact angle measurements of probe liquids on oxygen-based modified and unmodified PCL samples showed that surface hydrophilicity was increased significantly for 3-min plasma-modified PCL samples. The contact angle on unmodified PCL surface was 58° (±1) with ultrapure water, 34° (±1) with diiodomethane, and 71° (±3) with glycerol, whereas that for 3-min plasma-modified samples were 41° (±1), 38° (±1), and 31° (±2) for ultrapure water, glycerol, and diiodomethane, respectively. The ± represent standard deviation with $n=4$ for each plasma modification time and for each probe liquids.

The contact angle measurements were used to determine the solid surface energy of modified and unmodified PCL. Owens-Wendt's method was used to calculate total surface energy (σ_s) of PCL and its polar (σ_s^P) and dispersive (σ_s^D) components before and after plasma modification [25]. In Fig. 6.6, the variation in the total, polar, and dispersive solid surface energy of PCL with plasma modification are given. The total surface energy of PCL increased from 39 mN/m for unmodified to 51 mN/m for 3-min plasma-modified PCL samples. After plasma modification, there was no difference in the dispersive energies of PCL surface. However, the polar component increased significantly with the modification time contributing in the increment of total solid surface energy.

6.4.2 Surface Roughness

Atomic force microscopy (AFM) was used to quantify the surface roughness on PCL samples. A Dimension 3100 AFM (Digital Instruemnts, USA) was used in tapping mode at ambient conditions. The scan size was 5 μm, and the samples were scanned at a frequency of 1 Hz. Nanoscope 5.12 software was used to determine the surface characteristics of a surface quantitavely from AFM image data. Root-mean-square

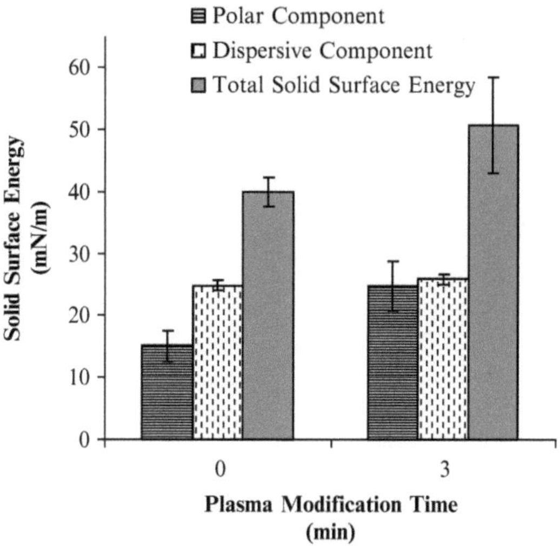

Fig. 6.6 The polar, dispersive and total surface energy (mN/m) of unmodified and 3-min oxygen-based plasma-modified PCL surface

roughness (RRMS), which is the standard deviation from the mean surface level of the image were measured by Nanocope software. In addition, phase AFM images of PCL film surface over a 5×5 μm square were plotted.

The RRMS roughness of PCL scaffold was increased from 41 ± 8 nm for unmodified PCL to 150 ± 12 nm for 3-min plasma-modified PCL scafffold. The AFM phase images (the three-dimensional) of modified and unmodified PCL surface are given in Fig. 6.7. While for unmodified PCL only very few features can be identified from the phase image (Fig. 6.7b), for 3-min modified sample the height of the features were increased and uneven feature distribution was observed (Fig. 6.7c). The results show that with the prolonged treatment time the mean surface roughness is increased in almost four times.

6.5 Cell–Scaffold Interaction

Scaffold of size 14×14×3 mm were seeded with primary fetal bovine osteoblast cells for a period of 21 days. The initial medium to maintain and proliferate the cells was prepared by mixing distilled H_2O, 13.5 g of DMEM powder mix (sigma, cat #D7777), 0.0059 g Ascorbic Acid (sigma, cat #A0278), 0.0588 g Gentamicin (sigma, cat #G3632), 3.7 g Sodium Bicarbonate (sigma, cat #S5761), 3.905 g HEPES Buffer (sigma, cat #H0763), and, 20 ml Antibiotic/Antimycotic solution (sigma, cat #A9909) into 1 L volume. The medium was then filtered through a 0.22-μm filter into sterilized bottles. 88 ml of FBS (sigma, cat #F2442) and 0.71 ml of 100× ITS (sigma, #I1884) per 500 ml medium were then added. For mineralization, 0.0783 g Calcium Chloride (sigma, cat #C7902), and 2.54 g beta-glycerol-phosphate (sigma, cat #G9891) supplements were added before filtering. The

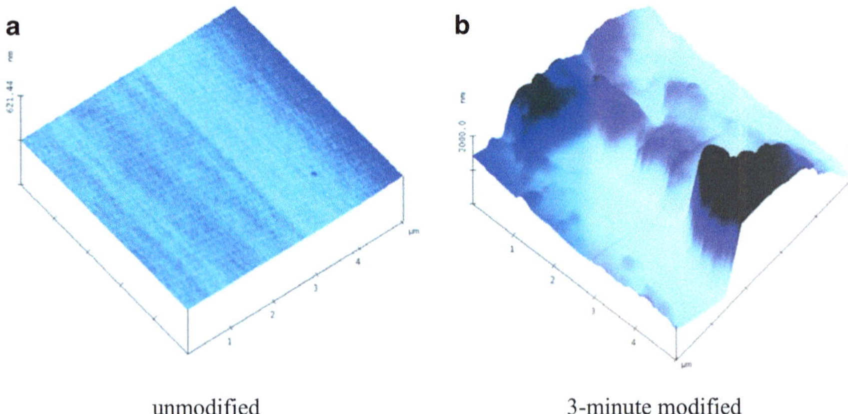

Fig. 6.7 Atomic force microscopy (AFM) phase images of oxygen plasma-treated PCL (**a**) unmodified, (**b**) 3-min modified

scaffolds were submerged in 70% ethanol for sterility and allowed to dry overnight. They were then washed with phosphate-buffered saline (PBS) and were soaked in 2 ml of medium for 1 h. Approximately 1.5×10^6 cells were seeded onto each scaffold. The cell–scaffold constructs were maintained in culture for 21 days.

6.5.1 Cell Viability and Proliferation

Alamar Blue assay, a flurometric indicator of cell metabolic activity, was performed to determine cell viability and proliferation. The cell–scaffold constructs were removed out of the culture plates on 3, 7, 11, 14, and 21 days. They were washed with PBS after aspirating the medium. They were then refed with 1.8 ml of medium and 0.2 ml of the Alamar Blue dye and allowed to incubate for 4 h. The resulting 2 ml solution was removed from the sample and the fluorescence was measured at room temperature on a plate reader (GENios, TECAN) using an excitation and emission wavelength of 520 and 590 nm, respectively. A cell number was obtained through a calibration curve determined by correlating a known cell number with the fluorescent intensity of the solution.

Figure 6.8 shows the results of Alamar Blue assay for PCL and PCL-HA scaffold. The cells did proliferate and their number increased over time until day 11 representing an active proliferation period. The data is presented as the average of the four samples ± standard deviation. There was a statistically significant difference ($P < 0.05$) in the cell numbers over the cultured time. However, there was no difference in the number of cells between PCL and PCL-HA. A slight decline was experienced between day 11 and 14. The decline in cell population could be due to

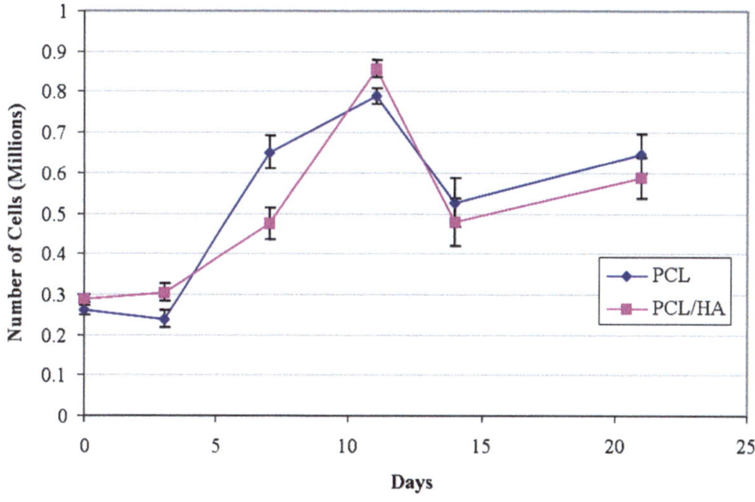

Fig. 6.8 Osteoblasts proliferation measured by Alamar Blue assay

the inability of Alamar Blue to react with cells trapped within the mineralization matrix after day 11 and/or due to slowing of cell proliferation as they migrate through the depth of the scaffold where nutrients are not as readily available. The increase between day 14 and 21 may indicate the active proliferation above the mineralization matrix and areas of the scaffold where calcification is less dense.

Figure 6.9 shows the results of Alamar Blue assay for PCL and 3-min plasma-modified PCL scaffold. The error bars represent ± standard deviation with $n=4$ for each group and each measurement day. Double asterisk indicates a significant difference in cell number between unmodified and plasma-modified scaffolds at the same time point. Compared to unmodified scaffolds, after a 7-day cell culture, the 3-min plasma-modified scaffolds had higher cell attachment efficiency (40%). As shown in Fig. 6.9, there was an upregulation of cell number on both modified and unmodified scaffolds, with the highest number at day 7. However, starting from day 14 the cell population began to level off and there were significantly fewer cells at day 21 compared to the first 2 weeks. We attribute this phenomenon on cell differentiation since it was proved the osteoblast cell proliferation decreases with starting of differentiation [26].

6.5.2 *Alkaline Phosphatase Activity*

Samples were removed of medium and washed twice with a buffer solution on 7, 14, and 21 days. The scaffolds were then submerged into 1 ml of 1% Triton ×100 solution for cell lysis. They were then centrifuged and the supernatant was used to calculate alkaline phosphatase (ALP) activity by p-nitro phenyl phosphate (p-NPP) method. Alkaline phosphatase catalyzes the cleavage of p-NPP to give

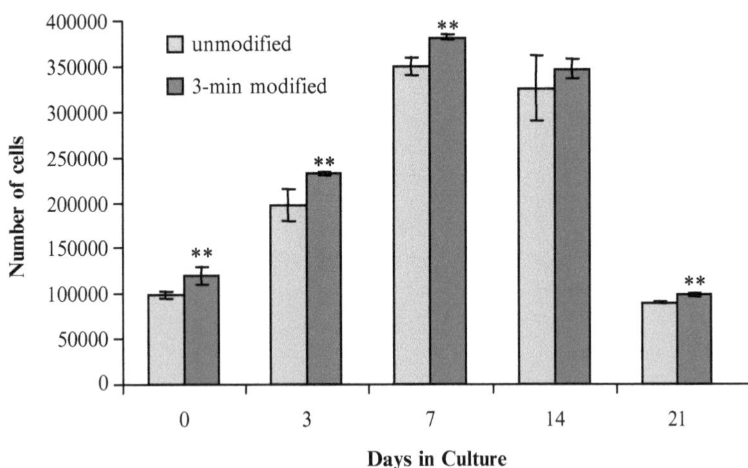

Fig. 6.9 The cell number on 3-min plasma-modified and unmodified PCL scaffolds upto 21 days

p-Nitrophenol and orthophosphate which develops a yellow color in the presence of a base. 0.5 ml of supernatant and 0.5 ml of diluted p-NPP (100 μl of p-NPP concentrate per 2 ml of 100 mM sodium bicarbonate/carbonate buffer, pH 10) were mixed and incubated for 45 min. The absorbance of this mixture was read at 405 nm. The absorbance was converted to the units of alkaline phosphatase per liter.

As seen from Fig. 6.10 an upregulation of alkaline phosphatase throughout the cultured time was observed, with the greatest increase within the first 14 days. PCL-HA scaffolds had higher ALP activity compared to PCL. Morphology and the extent of mineralization on the PCL surfaces were assessed using SEM after 21 days of culture. The SEM images show that mineralization was more over PCL-HA scaffolds compared to PCL.

6.5.3 *Cell Morphology by SEM*

The cell–scaffold constructs were removed from the culture media after 21 days. They were washed twice with 1× PBS and then fixed with 4% glutaraldehyde for 2 h. The constructs were then subjected to serial dilution of ethanol (20, 50, 70, 90, 100%), each 10 min for dehydration. They were refrigerated overnight at 4°C. The scaffolds were then coated with platinum and observed under SEM.

The positive effect of plasma modification on cell attachment could be confirmed by looking at the long-term mouse osteoblast cells adhesion on 3-min plasma-modified

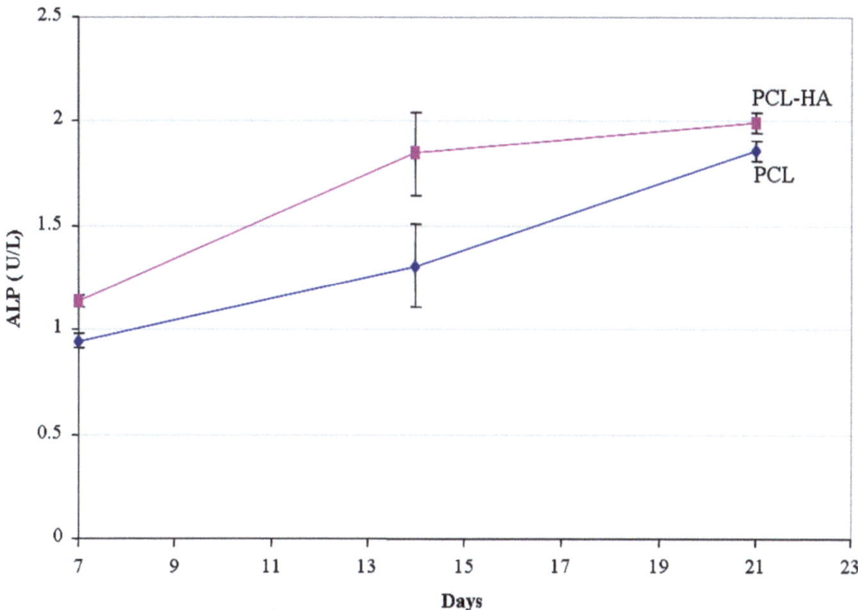

Fig. 6.10 Alkaline phosphates activity by osteoblasts cultured over PCL and PCL-HA scaffolds

PCL scaffolds by SEM. In Fig. 6.11, the SEM micrographs of 7F2 cells incubated on unmodified and modified PCL scaffolds at day 7 in low and high magnification were given. In low magnification (Fig. 6.11a), we could only observe individual cells scattered over the unmodified PCL scaffolds with a small number. However, Fig. 6.11b shows that the osteoblast cells on 3-min plasma-modified PCL scaffolds started to cover the struts of the scaffold with higher number. In high magnification SEM micrographs, the morphology of cells on scaffolds strut could be examined easily. For cell on 3-min plasma-modified PCL scaffolds (Fig. 6.11d) exhibited elongated morphology on scaffold strut with high degree of spreading. In contrast, on unmodified scaffolds osteoblast cells were hardly attached and preserved their round shape (Fig. 6.11c).

6.6 Statistical Analysis

The statistical significance was determined by analysis of variance (ANOVA) and Tukey post-hoc test at the significance level of less than 0.05 ($P<0.05$) using SPSS® version 14 for Windows® software package.

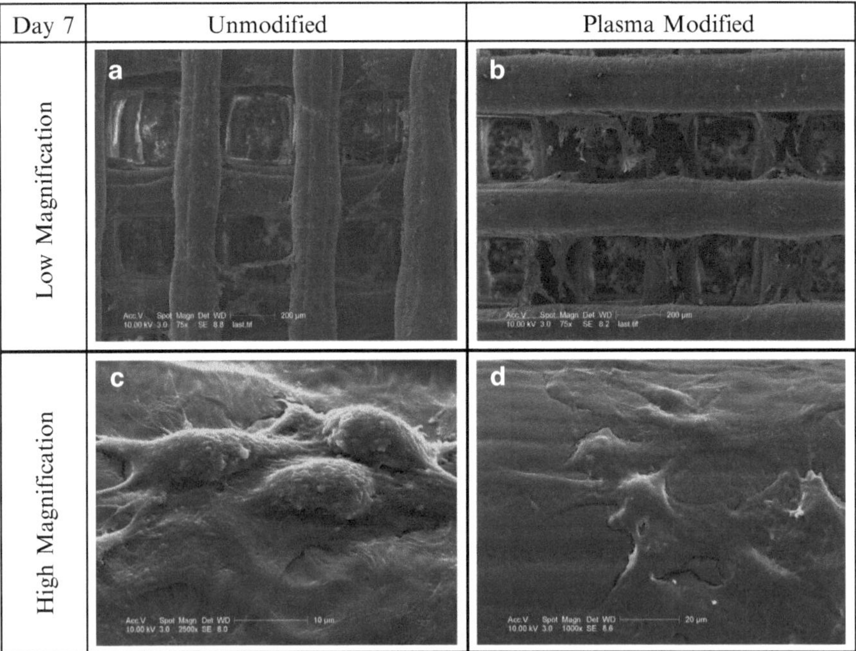

Fig. 6.11 The SEM micrographs of mouse osteoblast cells on unmodified and 3-min oxygen-based plasma-modified PCL scaffolds at day 7. Cells on (**a**) unmodified scaffolds with low magnification (**b**) 3-min plasma-modified scaffolds with low magnification (**c**) unmodified scaffolds with high magnification (**d**) 3-min plasma-modified scaffolds with high magnification

6.7 Discussion

The feasibility of using PED process for freeform fabrication of PCL and PCL-HA scaffolds with designed pattern was investigated. In contrast to the conventional FDM process that requires use of precursor filaments, the PED process directly extrudes scaffolding materials in its granulated form without the filament preparation step. This process is thus flexible, faster, and continuous with virtually no interruptions for a variety of biopolymer materials. The XYZ position system of the process can be used to precisely deposit the required material layer-by-layer with controlled architectures according to the design configuration.

Polycaprolactone is a biodegradable and biocompatible polymer with slow degradation rate adequate for bone tissue engineering applications. The surface hydrophobicity works against PCL when it comes to cell attachment [27–29]. Plasma surface modification and bioactive material inclusion can be used to improve the cell–material interaction of PCL.

To increase the bioactivity of PCL, we introduced HA to the material. HA is natural ceramic material found in the bones. We studied a composite material by melt blending PCL and HA to improve both mechanical and surface properties of the composite PCL-HA tissue scaffolds. In the reported study, 25% HA by weight was used to reinforce PCL. 60 and 70% porous scaffolds with 100% interconnectivity and pore size of 450 and 750 microns, respectively, were fabricated after optimizing the processing parameters of PED system. For fabrication of PCL-HA scaffolds PCL and HA were melt blended together to prepare a composite material. Our characterization using backscatter SEM images (Fig. 6.4) suggested that there were no visible clumps of HA and it was well-distributed. It is important that the HA particles used for the reinforcement should be well-distributed in the PCL matrix to avoid the nondesirable stress transfer in the scaffolds. Inclusion of HA significantly improved the mechanical properties of PCL (Fig. 6.5). Compressive modulus for 60% porous PCL scaffolds was 59 MPa and that of PCL-HA was 84 MPa, whereas for 70% porous PCL scaffolds it was 30 MPa and for PCL-HA was 76 MPa (Table 6.1). The porosity reduces the properties because of the lack of material in the scaffolds. The properties of composite scaffolds were similar to native properties of cancellous bone [23, 30] making them suitable for bone tissue scaffolds applications.

Besides HA inclusion, oxygen-plasma modification was used to elavate the quality of cell interaction through changing the physical and chemical properties of PCL. Our results indicate that oxygen plasma improved the surface hydrophilicity, energy, and roughness of PCL scaffold. The hydrophilicity and energy assessment were done by contact angle measurement and Owens-Wendt method proved that there was an increment in both surface hydrophilicity and total surface energy. The total solid surface energy was increased from 39 mN/m for unmodified to 51 mN/m for plasma-modified PCL samples. Although dispersive component of the solid surface energy did not significantly change with the plasma modification, the polar component increased significantly (Fig. 6.6). This proved that polar component was the dominant factor in the increment of total solid surface energy with the plasma modification. The main reason in increment of polar solid surface energy by oxygen-based plasma modification was that the increment in surface polarity by introduced oxygen containing polar functional groups on polymer surface [31]. Besides surface hydrophilicity and energy, plasma modification also increased the surface roughness of PCL scaffolds. The roughness was increased fourfold on 3-min plasma-modified PCL when compared to the unmodified one. The phase images of modified surface also proved the increased roughness by showing the introduced peaks and valleys on PCL surface (Fig. 6.7). It is belived that the change in surface morphology is the result from the ion bombardment and selective destruction of the polymer surface layer [32].

The cell–matrix interaction was carried out using fetal bovine osteoblasts. Alamar Blue assay was carried out for determining cell proliferation. It is a non-destructive assay wherein the dye Alamar Blue detects the metabolic activity of the cells. For HA incorporated scaffold, there was no statistical difference in cell proliferation between PCL and PCL-HA (Fig. 6.8). The cell numbers declined a bit on day 11 and 14. We believe that this might be due to the inability of dye to penetrate the scaffolds and react with the cells in the interior of scaffolds. The osteoblasts proliferation results in the formation of mineralized matrix. This was evident from the SEM images after 21 days (Figs. 6.12 and 6.13). This highly calcified matrix might have hindered Alamar Blue dye to go through the scaffolds. To demonstrate this we carried out a destructive assay for determining ALP activity using p-NPP method. The cell–scaffolds constructs were disrupted and cell lysis was prepared to perform the test. This allowed us to determine ALP from the cells without having any interference from the mineralized matrix surrounding the scaffolds. ALP is an enzyme produced by differentiating osteoblasts and is responsible for construction of bone matrix. A steady increase in the ALP activity was observed and the PCL-HA scaffolds had significantly higher ALP activity than PCL scaffolds (Fig. 6.10). The SEM images proved that PCL-HA scaffolds produced more mineralized matrix, compared to PCL scaffolds (Figs. 6.12 and 6.13). Moreover, the matrix was seen oriented along the pores making circular regions (Fig. 6.12a, b) similar to natural bone's Haversian systems [23]. These results demonstrated that the osteoblasts had improved differentiation on PCL-HA scaffolds compared to PCL. Experiments using different concentrations of HA, detailed in vivo and in vitro studies will be carried out in future and will be published in subsequent papers.

Beside the surface physicochemical properties of PCL, plasma modification could manipulate the cell–scaffold interaction. From the image of cell morphology via SEM and the proliferation assay, we can say that plasma modification of 3D PCL scaffold promote cell attachment,and proliferation. From SEM micrographs on day 7, we could say that cells on 3-min plasma-modified PCL scaffolds (Fig. 6.11) attached and elongated. 7F2 cell proliferation on PCL scaffolds was also improved by plasma modification (Fig. 6.9). Since most of the mammalian cells need to spread out on a substrate to proliferate, inadequate spreading due to poor adhesion can inhibit proliferation and differentiation [33]. Therefore, improved

Low magnification
High magnification
Top view
PCL
a
500μ
b
100μ
PCL Strut
Mineralized Matrix
Bottom view
PCL
c
500μ
d
100μ

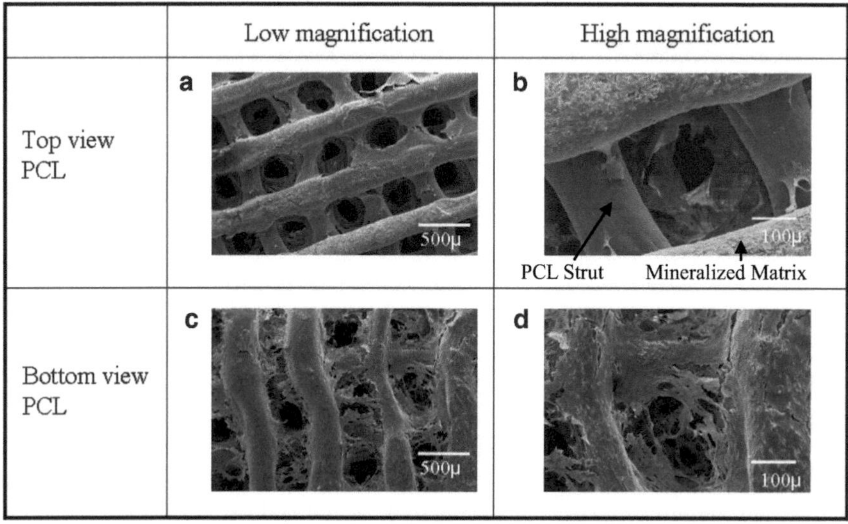

Fig. 6.12 Scanning electron microscopy images showing top (**a**, **b**) and bottom views (**c**, **d**) of mineralized matrix produced by osteoblasts cultured on PCL scaffolds after 21 days of culture

Low magnification
High magnification
Top view
PCL-HA
a
500μ
b
100μ
Mineralized Matrix
Bottom view
PCL-HA
c
500μ
d
100μ

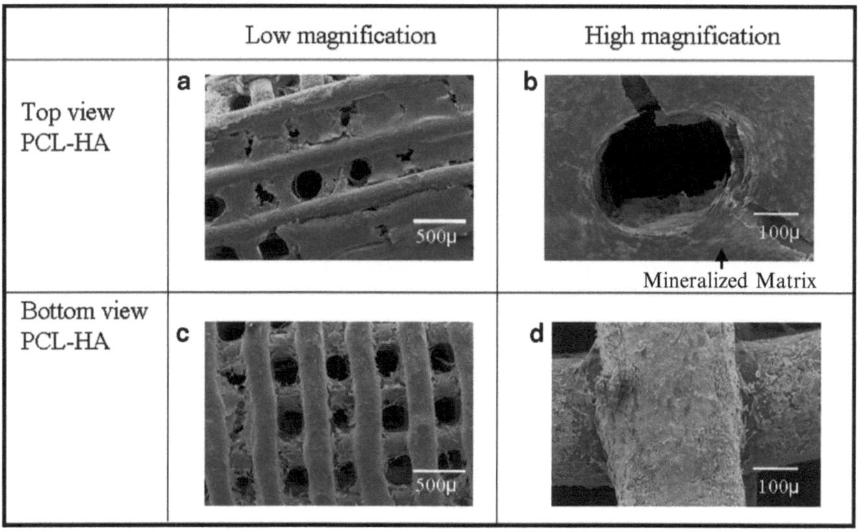

Fig. 6.13 Scanning electron microscopy images showing top (**a**, **b**) and bottom views (**c**, **d**) of mineralized matrix produced by osteoblasts cultured on PCL-HA scaffolds after 21 days of culture

cell adhesion could be the reason of increased cell proliferation in the first week of the culture. Compared to unmodified scaffolds, during the proliferation phase of 7F2 (after 7-day cell culture), the cell number on 3-min plasma-modified PCL scaffolds was higher than the number on unmodified one.

6.8 Conclusion

The ability to fabricate scaffolds of composite biomaterials using the PED system has been demonstrated. The scaffolds with controlled internal architectures were produced after optimizing the processing parameters. Results of the characterization demonstrated the capability of the PED fabrication process in manufacturing the scaffolds with controlled microstructure and pore size. The scaffolds were 60–70% porous with 100% interconnectivity. Success in scaffold guided tissue engineering requires a greater understanding of the cellular response to the constructed microenvironment. PED has the advantage of high precision on the micro scale as well as repeatability not available using more traditional scaffold manufacturing methods. This process opens opportunities for complex scaffold fabrication. Scaffolds of 60 and 70% porosity with pore sizes of 450 and 750 microns, respectively, were tested for their compressive properties. The composite PCL-HA scaffolds had significantly higher compressive modulus compared to PCL scaffolds. Cell–scaffolds interaction study demonstrated the biocompatibility of the process and material proving that the fabrication process has no adverse cytotoxic effect on the scaffolds. The plasma-modified PCL had improved cellular attachment and proliferation compared to unmodified PCL. In addition, the PCL-HA composite scaffolds had higher expression on ALP activity and showed more mineralization of matrix compared to PCL.

Acknowledgment Support from the National Science Foundation Grant No. 235342 "Computer-Aided Tissue Engineering" to this research is acknowledged.

References

1. Hollister SJ, Maddox RD, Taboas JM (2002) Optimal design and fabrication of scaffolds to mimic tissue properties and satisfy biological constraints. Biomaterials 23:4095–4103
2. Hutmacher DW (2000) Scaffolds in tissue engineering bone and cartilage. Biomaterials 21:2529–2543
3. Sun W, Lal P (2002) Recent development on computer aided tissue engineering – a review. Comput Methods Programs Biomed 67:85–103
4. Sun W, Darling A, Starly B, Nam J (2004) Computer-aided tissue engineering: overview, scope and challenges. Biotechnol Appl Biochem 39:29–47
5. Sun W, Starly B, Darling A, Gomez C (2004) Computer-aided tissue engineering: application to biomimetic modeling and design of tissue scaffolds. Biotechnol Appl Biochem 39:49–58

6. Taboas JM, Maddox RD, Krebsbach PH, Hollister SJ (2003) Indirect solid free form fabrication of local and global porous, biomimetic and composite 3D polymer-ceramic scaffolds. Biomaterials 24:181–194
7. Yang S, Leong K, Du Z, Chua C (2002) The design of scaffolds for use in tissue engineering. Part 2: rapid prototyping techniques. Tissue Eng 8(1):1–11
8. Simpson RL, Wiria FE, Amis AA, Chua CK, Leong KF, Hansen UN, Chandrasekaran M, Lee MW (2008) Development of a 95/5 poly(L-lactide-co-glycolide)/hydroxylapatite and β-tricalcium phosphate scaffold as bone replacement material via selective laser sintering. J Biomed Mater Res B Appl Biomater 84(1):17–25
9. Wiria FE, Leong KF, Chua CK, Liu Y (2007) Poly-e-caprolactone/hydroxyapatite for tissue engineering scaffolds fabrication via selective laser sintering. Acta Biomater 3(1):1–12
10. Tan KH, Chua CK, Leong KF, Cheah CM, Gui WS, Tan WS, Wiria FE (2005) Selective laser sintering of biocompatible polymers for applications in tissue engineering. Biomed Mater Eng 15(1–2):113–124
11. Chua CK, Leong KF, Tan KH, Wiria FE, Cheah CM (2004) Development of tissue scaffolds using selective laser sintering of polyvinyl alcohol/hydroxyapatite biocomposite for craniofacial and joint defects. J Mater Sci Mater Med 15(10):1113–1121
12. Venkataraman N (2000) The process-property-performance relationships of feedstock material used for fused deposition of ceramic (FDC). Rutgers University, New Brunswick
13. McNulty TF, Shanefield DJ, Danforth SC, Safari A (1999) Dispersion of lead zirconate titanate for fused deposition of ceramics. J Am Ceram Soc 8(7):1757–1760
14. Venkataraman N, Rangarajan S, Matthewson MJ, Harper B, Safari A, Danforth SC et al (2000) Feedstock material property – process relationship in fused deposition of ceramics (FDC). Rapid Prototyping J 5(4):244–252
15. Darling A, Sun W (2004) 3D microtomographic characterization of precision extruded poly-e-caprolactone tissue scaffolds. J Biomed Mater Res B Appl Biomater 70B(2):311–317
16. Engelberg I, Kohn J (1991) Physicomechanical properties of degradable polymers used in medical applications: a comparative study. Biomaterials 12(3):292–304
17. Hutmacher DW, Schantz T, Zein I, Ng KW, Teoh SH, Tan KC (2001) Mechanical properties and cell cultural response of polycaprolactone scaffolds designed and fabricated via fused deposition modeling. J Biomed Mater Res 55:203–216
18. Wang F, Shor L, Darling A, Khalil S, Sun W, Guceri S et al (2004) Precision extruding deposition and characterization of cellular poly-epsilon-caprolactone tissue scaffolds. Rapid Prototyping J 10(1):42–49
19. Li WJ, Tuli R, Okafor C, Derfoul A, Danielson KG, Hall DJ et al (2005) A three-dimensional nanofibrous scaffold for cartilage tissue engineering using human mesenchymal stem cells. Biomaterials 26(6):599–609
20. Karen JLB, Porter S, Kellam JF (2000) Biomaterial developments for bone tissue engineering. Biomaterials 21:2347–2359
21. Chen G, Zhou P, Mei N, Chen X, Shao Z, Pan L, Wu C (2004) J Materials Sci Mater Med 15:671
22. Vasilets VN, Kuznetsov AV, Sevast'yanov VI (2006) High energy. Chemistry 40(2):79
23. Gray H (2000) Anatomy of the human body. In: Lewis WH (ed) New York: BARTLEBY.COM
24. Bellini A (2002) Fused deposition of ceramics: a comprehensive experimental, analytical and computational study of material behavior, fabrication process and equipment design. Drexel University, Philadelphia
25. Yildirim ED, Ayan H, Vasilets VN, Fridman A, Guceri S, Sun W (2008) Plasma Process Polym 5:58–66
26. Owen TA et al (1990) Progressive development of the rat osteoblast phenotype in vitro: reciprocal relationships in expression of genes associated with osteoblast proliferation and differentiation during formation of the bone extracellular matrix. J Cell Physiol 143(3):420–430
27. Chen G, Zhpu P, Mei N, Chen X, Shao Z (2004) Silk fibroin modified porous poly-epsilon-caprolactone scaffold for human fibroblast culture in vitro. J Mater Sci Mater Med 15:671–677

28. Oyane A, Uchida M, Yokoyama Y, Choong C, Triffitt J, Ito A (2005) Simple surface modification of poly(epsilon-caprolactone) to induce its apatite-forming ability. J Biomed Mater Res A 75A(1):138–145
29. Chang G, Absolom D, Strong A, Stubley G, Zingg W (1988) Physical and hydrodynmaic factors affecting erythrocyte adhesion to polymer surfaces. J Biomed Mater Res 22:13–29
30. Rohl L, Larsen E, Linde F, Odgaard A, Jorgensen J (1991) Tensile and compressive properties of cancellous bone. J Biomech 24(12):1143–1149
31. Chim H et al (2003) Efficacy of glow discharge gas plasma treatment as a surface modification process for three-dimensional poly (D, L-lactide) scaffolds. J Biomed Mater Res A 65A(3):327–335
32. Grythe KF, Hansen FK (2006) Langmuir 22:6109
33. Wu S (1982) Polymer interface and adhesion. Marcell Dekker Inc, New York

Chapter 7
The Role of Technology in the Maxillofacial Prosthetic Setting

Betsy K. Davis, DMS, MS and Randy Emert

Abstract Medical imaging and rapid prototyping are viable tools which can be utilized in the process of creating extraoral prostheses. Successful implementation is a direct result of close collaboration between medical and engineering personnel. The use of medical imaging and rapid prototyping has the potential to reduce the cost and time in the fabrication of the wax patterns and could result in a more accurate morphologic result. This chapter describes the use of medical imaging and rapid prototyping used in the fabrication of an auricular wax pattern and its adaption to the clinical defect. The use of this technology results in a more symmetrical wax pattern and a significant time saving compared to sculpting in the traditional manner.

Maxillofacial Prosthetics is the subspecialty of prosthodontics in dentistry dealing with the prosthetic rehabilitation of head and neck cancer, trauma, and craniofacial patients. The specialty involves extraoral prostheses (Figs. 7.1–7.3) such as nasal, orbital, and auricular defects and intraoral prostheses such as obturators (Fig. 7.4), resection appliances, speech bulbs, and palatal lifts. For many years, the custom made prosthesis fabricated by hand was time intensive. The use of technology has the potential to transform the specialty in which surgical and prosthetic reconstruction are planned virtually, three-dimensional models are made for treatment planning, and the fabrication of custom made prostheses [1]. The use of medical imaging and rapid prototyping has been used more extensively with extraoral prostheses; whereas, the use of this technology in intraoral prostheses is at its infancy.

B.K. Davis, DMD, MS (✉)
Director, Maxillofacial Prosthodontic Clinic Associate Professor, Departments of Otolaryngology and Head & Neck Surgery and Oral & Maxillofacial Surgery Medical University of South Carolina, 135 Rutledge Avenue, Charleston, SC, 29425, USA
e-mail: davisb@musc.edu

R. Emert
Engineering Graphics, Clemson University, M-11 Holtzendorff Hall, Clemson, SC, 29634, USA

R. Narayan et al. (eds.), *Printed Biomaterials*, Biological and Medical Physics, Biomedical Engineering, DOI 10.1007/978-1-4419-1395-1_7,

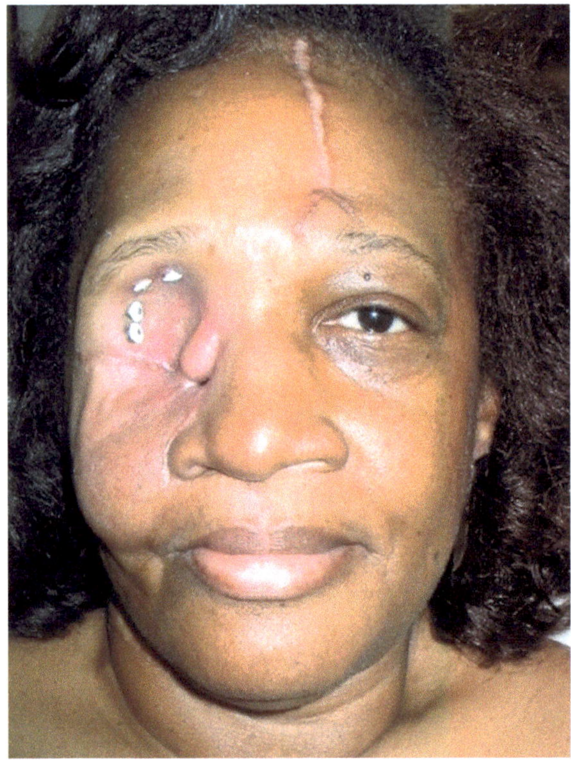

Fig. 7.1 An example of an orbital defect

7.1 Extraoral Prostheses

Medical imaging and rapid prototyping are viable tools which can be utilized in the process of creating an extraoral prostheses [2, 3]. Successful implementation is a direct result of close collaboration between medical and engineering personnel. The use of medical imaging and rapid prototyping has the potential to reduce the cost and time in the fabrication of an extraoral prostheses pattern and could result in a more accurate morphologic result. The use of this technology results in a more symmetrical wax pattern and a significant time savings compared to sculpting in the traditional manner.

This technology is applicable for auricular defects, orbital defecs, and nasal defects. Orbital and nasal defects are more likely due to cancer; whereas, auricular defects can be due to cancer or a craniofacial malformation. Auricular defects are the second most common craniofacial malformation and can be reconstructed either surgically or prosthetically (Fig. 7.5) [4]. Surgical reconstruction is complex due to the complex morphology of the ear. Prosthetic reconstruction results in a better morphologic result. However, prosthetic auricular reconstruction requires much

Fig. 7.2 An example of an orbital prosthesis

time and skill in the reproduction of an accurate anatomic wax pattern. The final prosthetic result is dependent on the operator's skill and artistry [4].

The use of medical imaging and rapid prototyping has the potential to reduce the cost and time in the fabrication of an auricular wax pattern and could result in a more accurate morphologic result. Part of the treatment planning process includes consultation with both the facial plastic and reconstructive surgeon to consider surgical reconstruction and the maxillofacial prosthodontist for an auricular prosthesis. If the patient elects for a prosthetic auricular reconstruction a CT scan is needed to obtain the wax pattern by the use of medical imaging and rapid prototyping.

A number of different methods have been used in the creation of the wax pattern [5–11]. Methods employed have been the traditional impression, scanned plaster cast, laser scanning, optical scanning, CT scan, and MRI scan. Auricular prosthetic fabrication begins with capturing the shape of the contralateral ear. For orbital and nasal defects, the pre-op CT scan is imaged to the post-op CT scan. Once the image of the contralateral ear has been obtained, a mirror image is created through a traditional process or through the use of special imaging software. If imaging software is used, the file is translated into an STL file for rapid prototyping or machine code file for milling. Rapid prototyping, in contrast to milling, is an additive process; whereas, milling is a subtractive process.

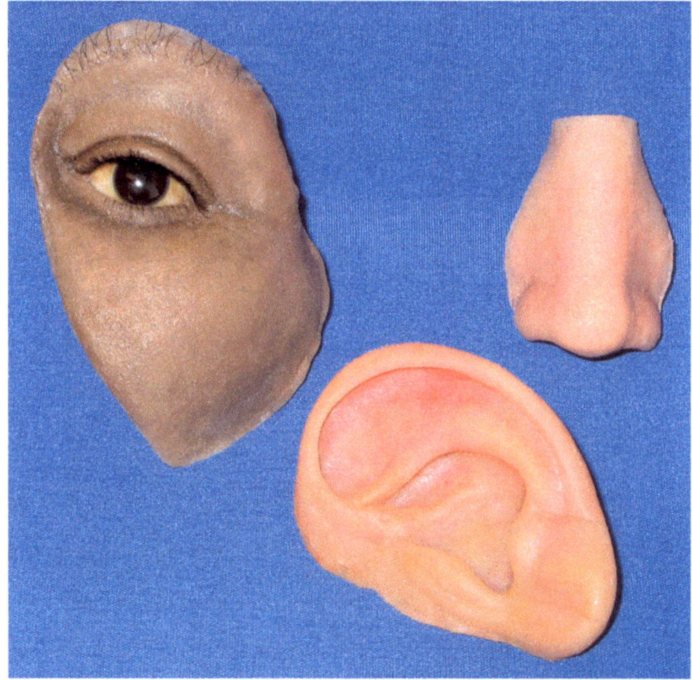

Fig. 7.3 Examples of orbital, nasal, and auricular prostheses

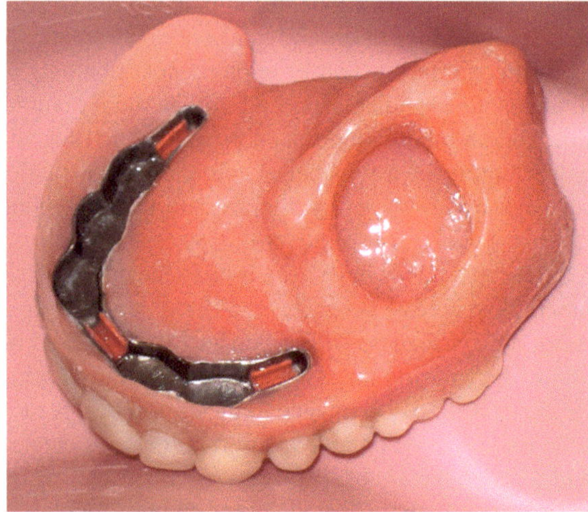

Fig. 7.4 An example of an obturator prosthesis in which the patient has lost part of her hard and soft palate

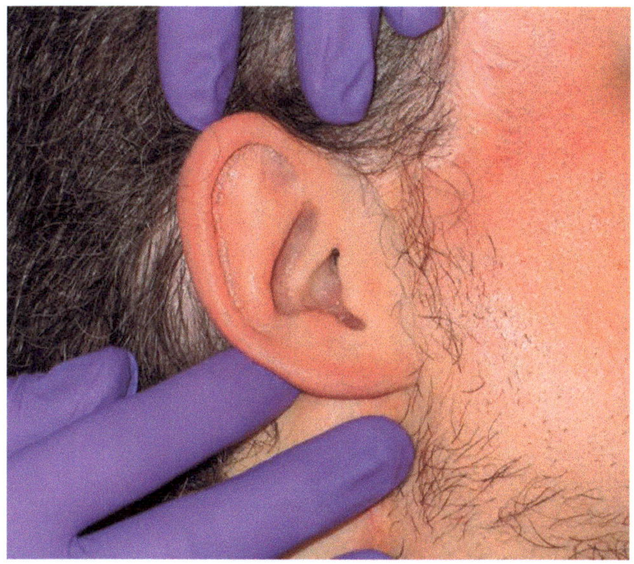

Fig. 7.5 An example of an auricular prosthesis

A CT scan along with medical imaging software (Mimics, Materialse Belgium) is used to create a pattern for the prostheses. Despite the immense capabilities of medical imaging systems, certain limitations are introduced in using these highly sophisticated technologies [12].

With CT technology, patients are exposed to relatively large doses of harmful radiation, and the spatial resolution encountered in soft tissue imaging is poor [8]. On the other hand, for many patients CT scans are readily available [10]. Therefore, it is not necessary to perform an additional procedure to capture data that already exists as with laser digitizing. The use of a CT scan eliminates the need for an impression visit to make an initial impression as is done with the traditional and scanning processes. Thus, the pain and discomfort associated with making an impression is eliminated. Both scanning and digitizing bring additional equipment and operations into the clinical setting that has the potential to detract from fitting and adjusting the prosthesis to the patient.

Mimics imaging software reads data in a standard DICOM format from many CT scanning equipment. The spatial imaging of soft tissue is more than sufficient to isolate the contour of the existing auricle. The process begins by isolating the existing auricle through thresholding, then performing standard boolean operations to remove excess material. The image is mirrored to create the pattern for the missing ear (Fig. 7.6). FreeForm software system (SensAble Technologies) has been used with digitizing to blend both the defect and remaining ear to create a virtual model of the required prosthesis; however, the model still needs to be adapted to

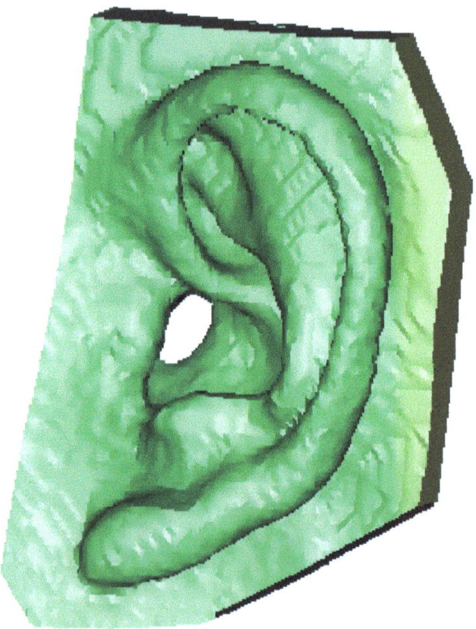

Fig. 7.6 Mirrored image of contralateral ear

the defect clinically [7]. There is a definite time tradeoff that exists between blending the model in the software and adapting the wax model in the clinic. Excess material remains on the model for the anaplastologist to work with during adaption of the wax pattern to the patient. The data is then translated into an STL file, which is standard for interfacing with rapid prototyping systems, 30 min is required to complete the task.

Milling and rapid prototyping have both been used as a step toward creating a wax auricular pattern. Milling is incapable of creating the contours to accurately represent the interior shape of the ear [7]. Milling also leaves material behind the ear which will later have to be removed manually. Several different rapid prototyping methods and machines have been used to create the wax pattern. The Z printer 310 (Z Corporation, Burlington, MA) creates an acrylic casting to cast the wax pattern [6], but an additional step is required to get to the wax pattern adapted to the defect. Fused deposition modeler Dimension (Stratasys, Eden Prairie, MN) can be used but a secondary operation is required to adapt the wax pattern to the defect. A Thermojet (3D Systems, Valencia, CA) is probably the best choice due to speed and material. The wax pattern takes 3 h to prototype (Fig. 7.7). The material Thermojet 88 (Remet Corporation, Utica, NY) is firm when handled but was pliable so the clinician could use traditional wax instruments to melt, form, and add additional material where needed. The melting temperature of the material is 90°C so the wax pattern can be boiled out following the traditional process to create the finished prosthesis. Although it takes approximately 3 h to produce a single model,

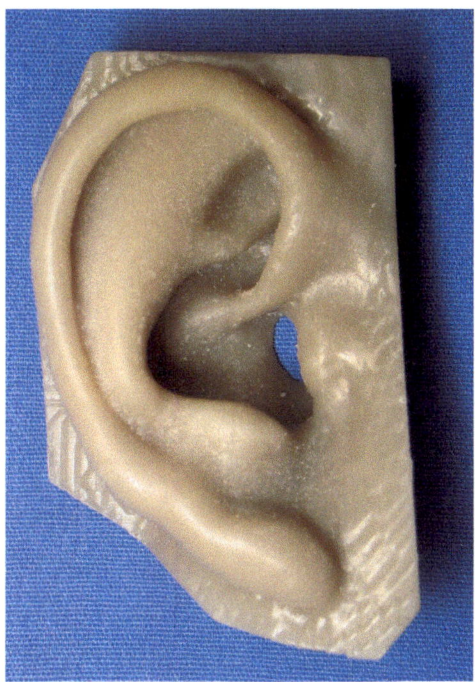

Fig. 7.7 Wax pattern made by rapid prototyping

a batch of 8–18 models, depending on the build orientation, can be produced within a similar production time.

After the creation of the wax pattern, the pattern is adapted to the patient. The pattern is then flasked and processed into silicone traditionally. The prosthesis is then painted by hand and delivered to the patient (Figs. 7.8–7.10). It is possible to take a digital skin color measurement with the Spectromatch system (Spectromatch, London, Great Britian) with the color formulation given to match the patient's skin tone [13].

Concurrent engineering, as it relates to surgical process planning (SPP), is a systematic approach that considers all elements of the patient's treatment from conception through functional outcomes, including quality, cost, schedule, and user requirements. The creation of a wax pattern is just one step in the overall process. Beginning with pre-operative planning surgeons and engineers, using 3D medical visualization software, can pre-operatively identify implant locations. The engineer prior to surgery can design a template and have it prototyped so the implants will be placed in the exact locations based on the pre-operative plan. It can also serve as a guide in the fabrication of the bar and suprastructure. DeCubber and Verdonck have developed software to design the bar in titanium with a layered manufacturing technique such as Selective Laser Sintering or Electron Beam Melting. Rather than using a template guide in the waxing of the bar, they recommended importing the

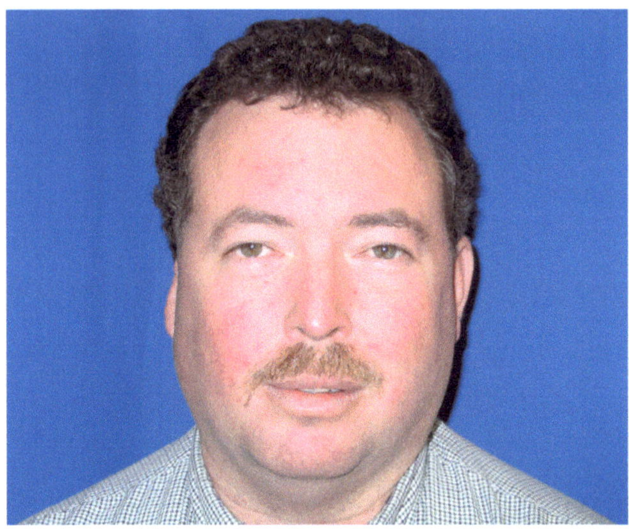

Fig. 7.8 Initial presentation of patient

Fig. 7.9 Completed prosthesis made with rapid prototyping

older CT data before resection to use as a guide virtually, so the bar is planned virtually [14]. The cast bar that is attached to the abutments may be designed directly after the design of the template and concurrently with the actual surgery. The wax pattern may then be created and ready for the patient on a first visit once

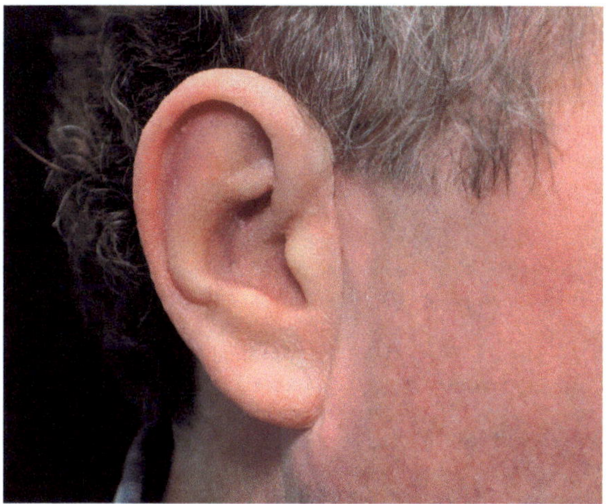

Fig. 7.10 Close-up view of prosthesis

the implants have osseointegrated. Using a concurrent process as opposed to a traditional linear one has the potential to greatly improve outcomes.

7.2 Intraoral Prostheses

The use of technology in dentistry and oral and maxillofacial surgery is growing rapidly with software supporting virtually planned surgery for osteotomies and implant placement. Prosthetic fabrication of intraoral prostheses with surgical stents is well-documented for conventional prosthodontics. However, the use of technology with intraoral prosthetic fabrication in maxillofacial prosthetics is still in its infancy. Okay and Buchbinder and Casey and Sullivan have reported on the use of osteotomy templates for the reconstructive surgeon in maxillary defects [1, 15].

7.3 Conclusion

Rapid prototyping technology and 3D medical modeling software are transforming the specialty of maxillofacial prosthetics. The use of this technology has resulted in an improvement in quality and a reduction in costs. It has proved advantageous for both patient and clinician. In the future, surgical and prosthetic reconstruction will be planned virtually with the use of 3D models and custom made prostheses by rapid prototyping.

References

1. Okay DJ, Buchbinder D (2008) 3D Planning for facial deformities: head and neck reconstruction and rehabilitation. In: 3rd International conference of advanced digital technology in head and neck reconstruction, Abstract K02, p 46
2. Gibson I, Cheung LK, Chow SP, Cheung WL, Beh SL, Savalani M, Lee SH (2006) The use of rapid prototyping to assist medical applications. Rapid Prototyping J 12:53–58
3. Hieu LC, Zlatov N, Vander Sloten J, Bohez E, Khanh L, Binh PH, Oris P, Toshev Y (2005) Medical rapid prototyping applications and methods. Assemb Autom 25:284–292
4. Jiao T, Zhang F, Huang X, Wang C (2004) Design and fabrication of auricular prostheses by CAD/CAM system. Int J Prosthodont 17:460–463
5. Al Mardini M, Ercoli C, Graser GN (2005) A technique to produce a mirror-image wax pattern of an ear using rapid prototyping technology. J Prosthet Dent 94:195–198
6. Ciocca L, Scotti R (2004) CAD-CAM generated ear cast by means of a laser scanner and rapid prototyping machine. J Prosthet Dent 92:591–595
7. Sykes LM, Parrott AM, Owen CP, Snaddon DR (2004) Applications of rapid prototyping technology in maxillofacial prosthetics. Int J Prosthodont 17:454–459
8. Cheah CM, Chua CK, Tan KH, Teo CK (2003) Integration of laser surface digitizing with CAD/CAM techniques for developing facial prostheses. Part 1: design and fabrication of prosthesis replicas. Int J Prosthodont 16:435–441
9. Cheah CM, Chua CK, Tan KH, Teo CK (2003) Integration of laser surface digitizing with CAD/CAM techniques for developing facial prostheses. Part 2: development of molding techniques for casting prosthetic parts. Int J Prosthodont 16:543–548
10. Coward TJ, Watson RM, Wilkinson IC (1999) Fabrication of a wax ear by rapid-process modeling using stereolithography. Int J Prosthodont 12:20–27
11. Webb PA (2000) A review of rapid prototyping (RP) techniques in the medical and biomedical sector. J Med Eng Technol 24:149–153
12. Guy C, Ffytche G (2000) An introduction to the principles of medical imaging. Imperial College Press, London
13. www.spectromatch.com
14. De Cubber J, Verdonck H (2008) Closing the circle (where advanced computer technology and the production of cranio facial epitheses meet). In: 3rd International conference of advanced digital technology in head and neck reconstruction, Abstract L03, p 59
15. Casey D, Sullivan M (2008) Stereolithographic models in treatment planning reconstruction after maxillectomy using bone graft and dental implants. In: 3rd International conference of advanced digital technology in head and neck reconstruction, Abstract P44, p 112

Index

H

I

T

U

V

W

If you have any concerns about our products,
you can contact us on
ProductSafety@springernature.com

In case Publisher is established outside the EU,
the EU authorized representative is:
Springer Nature Customer Service Center GmbH
Europaplatz 3, 69115 Heidelberg, Germany

Printed by Libri Plureos GmbH
in Hamburg, Germany